The Contemporary City

Series Editors
Ray Forrest
Lingnan University
Hong Kong

Richard Ronald
University of Amsterdam
Amsterdam, Noord-Holland, The Netherlands

In recent decades cities have been variously impacted by neoliberalism, economic crises, climate change, industrialization and post-industrialization and widening inequalities. So what is it like to live in these contemporary cities? What are the key drivers shaping cities and neighborhoods? To what extent are people being bound together or driven apart? How do these factors vary cross-culturally and cross nationally? This book series aims to explore the various aspects of the contemporary urban experience from a firmly interdisciplinary and international perspective. With editors based in Amsterdam and Hong Kong, the series is drawn on an axis between old and new cities in the West and East.

More information about this series at
http://www.palgrave.com/gp/series/14446

Ngai Ming Yip · Miguel Angel Martínez López · Xiaoyi Sun
Editors

Contested Cities and Urban Activism

palgrave
macmillan

Editors
Ngai Ming Yip
City University of Hong Kong
Kowloon, Hong Kong

Xiaoyi Sun
Fudan University
Shanghai, China

Miguel Angel Martínez López
Uppsala University
Uppsala, Sweden

The Contemporary City
ISBN 978-981-13-4671-2 ISBN 978-981-13-1730-9 (eBook)
https://doi.org/10.1007/978-981-13-1730-9

Cover credit: James Nesterwitz/Alamy Stock Photo

This Palgrave Macmillan imprint is published by the registered company Springer Nature Singapore Pte Ltd.
The registered company address is: 152 Beach Road, #21-01/04 Gateway East, Singapore 189721, Singapore

Acknowledgements

This book project started in a symposium held in March 2016 in City University of Hong Kong. The editors thank the participants in the symposium for their very active participation as well as the valuable and critical discussion on the papers that were presented there. The symposium was financially supported by the AXA Research Fund as well as the Urban Research Group of the Department of Public Policy, City University of Hong Kong. The Urban Research Group also provides the funding for the indexing work of this book. Xiaoyi Sun would like to express her gratitude to the AXA Research Fund which offered generous support for her postdoctoral fellowship at the City University of Hong Kong on her research project on "The Environmental Risk and Collective Action in Urban China". Miguel Martinez would like to acknowledge the support of the Research Grant Council of Hong Kong (project #11612016-CityU) which enabled him to co-edit this book as well as write Chapter 2. The editorial support of Sampson Law in the final editing of this book is also acknowledged.

Contents

Notes on Contributors

Ibán Díaz-Parra is Postdoctoral Researcher in the University of Seville. He has been previously affiliated to the University of Buenos Aires and to the National Autonomous University of Mexico. He has focused his academic research on urban studies, specifically studies in gentrification, socio-spatial segregation and housing, including housing movements and protest. His latest publications are as follows: (2016) Parra "Blurring the borders between old and new social movements: The M15 movement and the radical unions in Spain", *Mediterránean Politics*; (2015) Perspectivas del estudio de la gentrificación en México y América Latina (City of Mexico: UNAM); and (2015) "A back to the city movement by local government action: Gentrification in Spain and Latin America", *International Journal of Urban Science*, ORCID 0000-0002-4159-3338.

Federica Frazzetta is a Ph.D. student at the University of Catania. She graduated at the University of Catania in Political Science and International Relationships with a thesis on the No MUOS movement in Sicily. She specialised at the University of Trento in Sociology and Social Research with a thesis on unconventional political participation. Her interests of research are on collective action and social movements. Currently, she is working on the Italian right-wing galaxy.

Robert González is currently Professor and Researcher in the Area of Political Science and Public Administration at the Institute of Social Sciences and Humanities at the Universidad Autónoma del Estado de Hidalgo, Mexico. He was previously affiliated at the Universitat

Autònoma de Barcelona. In addition to his involvement in various social movements, he has various publications and executions of research projects on citizen participation, public policy, youth and social movements. His last publications are as follows: (2018) González, Robert; Díaz-Parra I., Martínez López M.A. "Squatted Social Centres and the Housing Question" *The Urban Politics of Squatters' Movements*. Palgrave Macmillan; (2017) Araiza, Alejandra and González, R. "La Investigación Activista Feminista. Un diálogo metodológico con los movimientos sociales" (Feminist Activist Researh. A methodological dialogue with social movements), Empiria. Revista de Metodología de Ciencias Sociales; and (2016) González, Robert and Araiza, A. "Feminismo y okupación en España. El caso de la Eskalera Karakola" (Feminism and Squatting in Spain. The case of the Eskalera Karakola"), Sociológica, ORCID 0000-0002-6166-5562.

Seon Young Lee holds her Ph.D. in Geography from King's College London and works as an Independent Researcher. She has published several journal articles and book chapters about gentrification and anti-gentrification movements in Korea. She has discussed social, physical and psychological aspects of gentrification as urban disaster and has analysed the contribution and limitation of cultural resistance in recent anti-gentrification movements in Korea.

Anqi Liu Having spent years working with migrant communities and witnessing the challenges of marginalised communities, the author opted for theoretical contemplation and especially focused on the role of social movements and intentional communities. She obtained an MA in Global Studies from Albert-Ludwigs-Universität Freiburg (Germany); Facultad Latinoamericana de Ciencias Sociales (Argentina); and Jawaharlal Nehru University (India), and her research interest includes global civil society, media and violence, deep ecology, etc. She now finds herself increasingly in farming and meditation that supplement structural approaches with a different sense of resilience.

Danny Marks is an Assistant Professor of Environmental Studies at the Department of Asian and International Studies of City University of Hong Kong. Prior to this position, he was a Postdoctoral Research Fellow with the Urban Climate Resilience in Southeast Asia project at the Munk School of Global Affairs of the University of Toronto. Dr. Marks has spent a number of years conducting research and working in Southeast Asia, particularly in the fields of climate change adaptation and environmental governance. He has worked for a number of organisations in the region, including the World Bank's East Asia and Pacific Governance Hub, the Rockefeller Foundation, ActionAid and the NGO Forum on Cambodia. Dr. Marks completed his Ph.D. dissertation, An Urban Political Ecology of the 2011 Bangkok Floods, at the University of Sydney. He received his MA in International Affairs from the Johns Hopkins School of Advanced International Studies. He has published on climate change governance, disaster risk reduction and Thai domestic politics in numerous academic journals, blogs and newspapers.

Miguel Angel Martínez López is currently affiliated to the IBF (Institute for Housing and Urban Research), Uppsala University (Sweden), as Professor of Housing and Urban Sociology. His academic background is rooted in the field of urban studies and especially in the intersection between urban sociology, political science, social movements and urban planning. Regarding the research on urban movements, struggles and activism, he has published extensively about the squatters' movements in Europe (last edited book: *The Urban Politics of Squatters' Movements* (ed.) New York: Palgrave Macmillan). Since 2011, following the "occupy movement" in Madrid (Spain), he investigated the connections between squatting for housing, anti-neoliberal mobilisations and "new-left" municipalism in Spain. Three main publications came out of this research: "Bitter wins or a long-distance race? Social and political outcomes of the Spanish housing movement", Housing Studies; "Converging Movements: Occupations of Squares and Buildings" (co-authored with A. García) in B. Tejerina and I. Perugorria (Eds.) *Crisis and Social Mobilization in Contemporary Spain. The 15M Movement*. Abingdon: Routledge, 2018; and "Between autonomy and hybridity: Urban struggles within the 15M movement in Madrid" in Margit Mayer, Catharina Thörn and Håkan Thörn (Eds.) *Urban Uprisings: Challenging the Neoliberal City in Europe*. London: Palgrave, 2016. During the years of his stay in Hong Kong, he also conducted research on urban and political processes in China and South-East Asian countries. A forthcoming publication in the journal *Social Justice* is "Street Occupations, Neglected Democracy and Contested Neoliberalism in Hong Kong". Most of his works are publicly available at www.miguelangelmartinez.net.

Inés Morales-Bernardos is a forest engineer and specialist in agroecology and organic farming. Currently, she is conducting a Ph.D. at the Institute of Sociology and Peasant Studies (ISEC), University of Córdoba, Spain. She researches urban movements and their politics of food provisioning in the peripheries, with case studies in Madrid, Athens, Lisbon and Naples. She has written several papers around food movements, and currently, she is co-editing the book *The Right to the City and Social Ecology—Towards Ecological and Democratic Cities*, Black Rose Books. As an activist, she has been involved in several autonomous urban and rural movements.

Gianni Piazza holds a Ph.D. in Political Science at the University of Florence and is an Associate Professor of Political Sociology at the University of Catania. He is the Associate Editor of the scientific journal Partecipazione e Conflitto and has published books and journal articles on local government and politics, public policy analysis, social movements, local and environmental conflicts, squatting social centres. He is the author of *La città degli affari* (1994) and *Sindaci e politiche in Sicilia* (1998), the co-author of *Politiche e partecipazione* (2004), *Voices of the Valley, Voices of the Straits* (2008), *Le ragioni del no* (2008), the editor of *Il movimento delle occupazioni di squat e centri sociali in Europa* (2012) and the co-editor of *Alla ricerca dell'Onda* (2010).

Sonia Roitman is a Senior Lecturer in Development Planning at The University of Queensland, Australia. Her work addresses the manifestation of inequalities in cities, in particular how community groups organise themselves and how the city reproduces inequalities through real estate projects and unjust policy frameworks. Her current research locations are Argentina, Indonesia and Uganda. She is a board member of the Research Committee of the Sociology of Urban and Regional Development, International Sociological Association (RC21) and also participates in the UNI-Habitat partnership.

Tom Slater is Reader in Urban Geography at the University of Edinburgh. He has research interests in the institutional arrangements producing and reinforcing urban inequalities, and in the ways in which marginalised urban dwellers organise against injustices visited upon them. He has written extensively on gentrification (notably the co-authored books, Gentrification, 2008 and *The Gentrification Reader*, 2010), displacement from urban space, territorial stigmatisation (the co-authored book *The Sociology of Stigma*, 2018), welfare reform, and social movements. Since 2010, he has delivered lectures in 19 different countries on these issues, and his work has been translated into 10 different languages and circulates widely to inform struggles for urban social justice. For more information, see http://www.geos.ed.ac.uk/homes/tslater.

Xiaoyi Sun is an Assistant Professor of Urban Studies in the School of International Relations and Public Affairs, Fudan University (Shanghai, China). She obtained her Ph.D. degree at the Department of Public Policy of City University of Hong Kong. She conducted her post-doc research project on "Environmental Risks and Collective Action in Urban China" sponsored by the AXA Research Fund. Her research focuses on environmental politics, urban sociology and neighbourhood governance in China. Her articles have recently appeared in journals including *Journal of Urban Affairs*, *Urban Affairs Review*, *Journal of Contemporary China*, *China: An International Journal*, etc.

Hongze Tan is an Assistant Professor of Sociology at Nankai University (China). He received his Ph.D. in public policy from the City University of Hong Kong. His research interest focuses on the field of urban studies, especially in the intersection between urban sociology, urban activism, sustainable development and mobilities studies. Regarding the research on urban sustainable development, particularly the issue of urban water source protection and governance, he has a series of academic publications in both English and Chinese. From 2014 onwards, he began to investigate urban mobility and active-transportation development. He works on exploring the historical and current changes regarding cycling governance in urban China and the interactions among the involved governmental and non-governmental actors. Based on this research, he finished his doctoral thesis and several forthcoming journal articles and book sections. Most of his works are available at: https://www.researchgate.net/profile/Hongze_Tan2.

Ngai Ming Yip is a Professor of Housing and Urban Studies in the Department of Public Policy, City University of Hong Kong. He worked as a social worker on youth and community development for ten years before joining the academia. He researches on urban and housing issues in Hong Kong, Mainland China, Vietnam and other Asian countries. He has been researching on housing activism and neighbourhood governance in urban China and has been working on a number of related projects. In the area of housing activism and urban struggle, he has published an edited volume on *Neighbourhood Governance* (2014, Edward Elgar); Internet and Activism in Urban China: A Case Study of Protests in Xiamen and Panyu (2012, *Journal of Comparative Asian Development*); Homeowners United—The Attempt to Create Lateral Networks of Homeowners' Associations in Urban China (2011, *Journal of Contemporary China*); Infrastructural power and neighbourhood governance: Transformation of Residents' Committees in Shanghai in the 1980s (2017, *China: An International Journal*); and Dynamic Political Opportunities and Environmental Forces Linking up: A Case Study of Anti-PX Contention in Kunming (2017, *Journal of Contemporary China*).

List of Figures

List of Tables

PART I

Introduction

CHAPTER 1

Introductory Remarks and Overview

Ngai Ming Yip, Miguel Angel Martínez López and Xiaoyi Sun

The notion of 'urban activism' holds an ambiguous status in both the fields of social movements and urban studies. It usually conflates the meanings of 'urban movements' and all sorts of activist practices that take place in cities. 'Urban movements' is the conventional expression to capture sustained mobilisations and protests that challenge consolidated power structures in relation to the production and transformation of urban spaces. When these collective actions are neither lasting over time nor sufficiently challenging the rulers and managers over the territory, it seems convenient to just designate them broadly as 'activism'. Therefore, we argue that urban activism occurs within specific organisations not much coordinated with others, and when it is focused on single-issue

N. M. Yip (✉)
City University of Hong Kong, Kowloon Tong, Hong Kong
e-mail: sayip@cityu.edu.hk

M. A. Martínez López
Uppsala University, Uppsala, Sweden
e-mail: miguel.martinez@ibf.uu.se

X. Sun
Fudan University, Shanghai, China
e-mail: sunxiaoyi@fudan.edu.cn

N. M. Yip et al. (eds.),
Contested Cities and Urban Activism, The Contemporary City,
https://doi.org/10.1007/978-981-13-1730-9_1

demands and campaigns with limited duration and capacity to alter the deep roots of urban politics. Activists are engaged participants in collective action as members of disparate groups who can turn into contributors to larger movements under certain circumstances. We elaborate on this distinction more accurately.

Urban activism and movements have undergone waves of transformation in the past few decades. Earlier research tends to focus on large scale nationally or even internationally orchestrated campaigns which were often elaborated and 'institutionally-heavy' with reliance on professional organisers and involved in situ participation of supporters. Yet this has been overshadowed by the emergence (or revival, according to some) of 'new social movements' since the 1970s which were often fragmented and 'institutionally-thin' with non-materialistic demands around issues like identity, gender, ecology or idealistic mottos like peace (Roth 2000). However, more recently, 'old problems' like employment, housing and urban planning returned to the scene. The contradictions created by late capitalism, the retreat of the welfare state, and the advancement of neoliberalism are believed to be the underlying driving forces. Shared concern on the impacts of global events like climate change, impacts caused by the global flows of capital have underpinned many protests over the recent decades (Hamel 2000). Isolated actions are also being linked to each other by resources and mediation from international NGOs and movement organisations. Hence, as Harvey (2008) postulates, even localised urban struggles have to be approached from a global scale as global financial capitals are dominating most urbanisation processes.

In contrast to the global north to which most of the studies on urban activism make reference, the form and process of struggles in the global south take very different shape despite sharing the same global threats, like the environmental crisis and the proliferation of financial powers (as well as local government acting as their agents or in collusion with them). High in the agendas of urban activism in the global south are collective consumption issues like clean water, basic shelter, actions against displacement and better conditions for street vendors. Such localised actions are cut off from the international movement networks and are little known outside of their action localities (Mayer 2009). Indeed, integration of the actions in cities in the global south with the specificity of the local sociopolitical environments is often more pivotal than their connection with global movement networks or international NGOs.

Hence, local singularities remain as a crucial dimension in the study of urban activism and movements, despite their tensions with global scales or even the transnational diffusion of protest repertoires.

The shaping of the local specificity of urban activism goes beyond the immediate local context but is connected to the social, political and cultural environments at various levels—local, regional, national and even global. In this respect, as most empirical cases have been reported from the global north within the tandem formed by capitalism and liberal democracy, bias in the analysis appears to be inevitable. The assumptions that 'civil rights are protected, the press is free, courts and legislatures are independent of the executive, and mechanisms for regular transfers of political power are institutionalized' (Osa and Schock 2007) simply cannot be taken for granted for many countries in the global south. The recent erosion of those pillars in countries of the global north such as Russia and Hungary, or in developing countries (Turkey, Brazil, etc.), warns us about the constraints of national and local contexts for the expression of grass-roots claim-making.

Situations are even worse in transitional economies under semi-authoritarian rule, like China and Vietnam, in which 'repression or the threat of repression is omnipresent … the risk of associating with regime opponents and advocating policies not sanctioned by authorities is costly, and resources that may be used to oppose the regime are difficult to acquire' (Osa and Schock 2007, p. 124). Collective actions of protest are still being perceived as a threat to social stability and hence various forms of activism are censored by state-controlled conventional media or being filtered in the new online media (for instance, by the Great Firewall in China). Help or support from international networks may be also counterproductive because it attracts the attention of the state and ends up in even more extensive suppression.

Even in countries in which western-like liberal democracy is imitated or been newly set up (e.g. Central and Eastern Europe), political resources and channels of redress may have been monopolised by the elites and the media in addition to non-independent judiciaries. Equally notable are countries in North Asia. Despite their capitalist system is arguably well advanced, their developmental approach to economic growth put the role of the state in a dominant position which is further complicated by the interweave of social factions within closely knitted political interests.

Activism, Movements and Urban Politics: Theoretical Concerns

Academic definitions on urban movements rarely distinguish various forms of activism as their key components. We argue that although all forms of urban activism nurture urban movements, the latter generally imply more intense contentious politics than the former. Activism also refers to less coordination between disparate groups and less tendency to expand their social networks of alliance, support and action. We thus contend that the above are differences of degree more than essential distinct features. Movements cannot exist without activists, but many activist expressions are not able to shape larger movements. Overlaps may also occur between them and other forms of urban politics we cannot examine here (from political parties and elections to charities, sport clubs and hidden or dispersed forms of resistance to authority: Scott 1990).

The meaning of the 'urban' holds obvious spatial connotations. Local places and cities are the location for manifold protests but, for us, only those involving the configuration, production and transformation of spaces within a local scope (from neighbourhood sites to the metropolitan scale) should be considered 'urban'. This stance also excludes spatial determinism because we take for granted that urban spaces are, mainly, the result of social (as well as political, economic and cultural) processes which make and reproduce a dominant mode of production and consumption (Harvey 1996, Ch. 9). However, as mentioned above, the transnational nature in many locally bounded urban struggles has often been neglected (Mayer and Boudreau 2012, p. 284). Many movements have to deal with the politics of scales when authorities operate rescaling processes of resources, policies and decision-making processes, which oblige activists to 'multi-scalar strategies' (Martínez 2018a; Nicholls et al. 2013, p. 9).

Based on these premises, we define 'urban activism' as the social practices of protest and claim-making about urban affairs within specific economic and political contexts—usually, in short, a capitalist society (Heatland and Goodwin 2013; Pickvance 1995, p. 198; Pruijt 2007). The more they persist and tackle the power structures of cities, the more they can scale up to the category of urban movements. Hence, when activism is fully channelled through state institutions in the form of interest groups, it rarely raises the stakes of political contention between the citizenry and the political-economic establishment. The more institutionalised the demands from below and responsive the authorities, the less need for activism. Sometimes the opponents of activists are not the state officials and rulers but private companies and managers, and even other social groups according to their residence, ethnic status and stigmas (Tyler and Slater 2018).

A brief account of two major theoretical approaches may further illuminate the framework that guides our inquiries on the topic of urban activism and movements. Firstly, the 'contentious politics' approach refers to 'people struggling with each other over which political program will prevail' (Tilly and Tarrow 2007, p. 2). This involves interactions between actors, claim-making bearing on someone else's interests, some forms of social organisation or coordination, and the direct or indirect involvement of governments. Not every expression of collective action is thus contentious. Some may remain within the walls of the state bureaucracy and representative democracy (voting, lobbying, running for office and so on) while others may scale up to social turmoil, uprisings and riots (or even wars). Urban movements and urban activism usually fall in the midst of those two poles. Tilly and Tarrow define social movements according to their routinised continuity in terms of claim-making campaigns, public performances (marches, petitions, meetings, etc.) and 'public displays of worthiness, unity, numbers and commitment by such means as wearing colors, marching in disciplined ranks, sporting badges that advertise the cause, displaying signs, chanting slogans, and picketing public buildings' (Tilly and Tarrow 2007, p. 8). Crucially, for them and us, activists comprise the 'social movement bases' besides organisational and cultural resources. However, if these social movement bases do not give birth to 'political identities' (a similar argument is held by Andretta et al. 2015, p. 203) and, simultaneously, do not turn into sustained 'movement campaigns', activism becomes only a kind of 'mini-movement' (our term) or one-time protests that can hardly threaten the legitimacy of the power holders.

Despite the dominant realm of state politics and policies, social movements are specialised in exploring and promoting repertoires of action beyond state institutions. Overall, political conditions either facilitate or hinder activism. They may operate not only at the local but also, very often, at the regional, national and even international-global scales. Hence, Tilly and Tarrow point out 'scale shift' as one of the main processes that researchers must reconstruct and explain. This concept is particularly useful in order to incorporate an urban or spatial view into the analysis of political mobilisation. According to them, scale shift means both the diffusion of contention and the coordination of collective action at higher (upward) or lower (downward) levels. Although most activism is locally initiated, it can shift venue 'to sites where contention may be more or less successful, and can threaten other actors or entire regimes' (Tilly and Tarrow 2007, p. 95). Instead of just focusing on the venue or location of protest, we can expand the concept to the spatial goods or policies (e.g. housing, or education and health facilities) addressed by activists at different state instances when their claims are diffused by emulation, territorial expansion of the main organisation and appeals to social groups from other cities, and subject to the same deprivation framed by the early risers (see also Martin and Miller 2003, p. 148; Nicholls et al. 2013, pp. 8–10).

Only on a marginal note, Tilly and Tarrow mention that activists' claims might be on other targets such as 'owners of property, religious functionaries, and others whose actions (or failures to act) significantly affect the welfare of many people' (Tilly and Tarrow 2007, p. 119). To complete the picture, we might add various social groups and hegemonic worldviews (or cultural codes) as the direct targets of activists (Hamel et al. 2000; Dikec and Swyngedouw 2017, p. 4). Another criticism concerns the role of social movement organisations. They can often be replaced by loose social networks and the mobilisation of different supporters and allies according to class, gender, cultural, ideological and spatial allegiances (Goodwin and Jasper 2004). Strong and durable campaigns, struggles and political identities seem more significant than the nature of movement organisations—although their mobilised resources may make a difference. Finally, a tight connection of social movements with their most significant context, the development of capitalism and urbanisation (Mayer and Boudreau 2012, p. 279; Hetland and Goodwin 2013; Rossi 2017) is usually missing in the contentious-political process approach.

A second approach that nurtures our views on movements builds upon the tradition of critical political economy. In the 1970s and 1980s, Castells defined 'urban movements' as those able to both transform the whole urban system of planning and management, and to challenge the reproduction of capitalist relationships (Castells 1972, pp. 321–323). For him, less challenging forms of protest and community participation would match what we designate as 'urban activism'. This distinction implied a hierarchy between them (revolutionary movements vs.

reformist citizen participation) which dismissed the aforementioned overlaps (movements encompassing activism, protest and participation) and the fact that most urban movements do not necessarily defy the heart of capitalism. However, Castells offered two valuable premises for the analysis of urban movements: (1) the articulation of economic, political, social and ideological dimensions of urban movements and activism within the hegemony of capitalism; (2) the space as a dimension of class struggles when turned into the field of 'collective consumption'—the spatial means and public services necessary for the reproduction of the labour force (Saunders 1983, pp. 113–127). These tenets still ruled in Castells' next twist (1983) where the examined cases clearly showed low-key anti-capitalist activism. This result led him to replace the target of urban movements: instead of being able to change the urban system and the capitalist society, they just tackle the 'urban meaning'. This 'urban meaning' still refers to class and socioeconomic struggles, but it centrally incorporates a historical and cultural skeleton (Castells 1983, p. 302). In addition to class, he introduced gender, ethnicity and residents' aspirations to control their own local environment, facilities and administration as key components of urban movements/activism (Castells 1983, p. 291).

Castells was criticised for endorsing urban movements with a high capacity to produce social change without taking into account the performance of state authorities, capitalists' actions and even institutional means of claim-making such as voting, petitions and lobbying (Pickvance 1975, pp. 32–39). His theoretical models did not integrate well the structural contexts in which every movement exists such as 'the coexistence with a broader political movement, the presence or absence of political parties, and state structure and government policy' (Pickvance 1985, p. 35). In addition, political opportunities are usually articulated with other contextual features regarding the urbanisation process, the economic cycles and cultural trends about, for example, the quality of urban life (Pickvance 1985, pp. 40–44; Fainstein and Hirst 1995, p. 198; Lowe 1986, pp. 152–186; Marcuse 2002).

It is also worth noting that other scholars engaged with this approach from alternative angles. Mayer (1993, p. 149), for example, showed how empowered urban movements of the 1970s in West Germany such as rent strikes, squatting and massive demonstrations against urban renewal became increasingly associated with other social movements (such as the women and the environmental ones) and that eventually led to electoral alliances.

Later on, many community groups were incorporated in municipal programmes and moderated their demands. However, instead of an inevitable path to their institutionalisation and defeat, urban movements became more diverse, fragmented and occasionally able to get mobilised again. In the late 1990s and the early 2000s, the anti-globalisation wave brought many urban movements into broader coalitions in order to oppose privatisation and welfare state dismantling (Mayer 2006, p. 203; Della Porta 2015). Hence, the new context of globalisation transformed urban politics due to city-branding strategies, construction of mega-projects, shrinking municipal budgets, outsourcing of services and competition for attracting capital investment (Mayer 2016).

Another alternative strand of analysis was provided by pioneering Harvey's (1973, 1982) insights into the capitalist city. Instead of focusing on 'collective consumption'—i.e. the spatial conditions for the reproduction of the labour-power—Harvey promoted the notion of urban space as the reproduction of capital, the material basis for producing rents and profit by means of private property and speculation. 'The city in Harvey's analysis assumes a much more dynamic significance. It is a 'productive' rather than reproductive instrument within capitalism' (Merrifield 2014, p. 20). According to Castells, urban movements tend to contentiously engage with the state provision and management of collective consumption. As for Harvey, urban movements would not be necessarily attached to the specific local places where they live and protest because their goal is, if they take a progressive stance, to interrupt the circulation of capital wherever the space it takes over. Movements should thus be studied in their connections to the context of global flows of capital and the urban landscapes which are configured, occupied and manipulated according to the capitalists' interests. As Merrifield contends, since the neoliberal turn initiated in the 1970s, even social democratic municipal governments succumbed to state encroachment, divested from collective consumption budgets, privatised and outsourced most of their services. In turn, they started subsidising capital in order to befriend spatially attached global investments landing in their cities (Merrifield 2014, p. 19). Many urban movements reacted to the core of these crucial processes, while others behave within this context without even attempting to challenge it.

Unlike Castells, Harvey was not initially focused on urban movements' research. This changed over the years when he recalled the notion of 'militant particularism' to indicate that there are many particular and local grass-roots struggles whose 'interests, objectives and organisational forms are fragmented, multiple and of varying intensity' (Harvey 2001, p. 190). This implicitly invites to ask, in our terms, how urban activism shifts into urban movements or 'broader [institutional] politics'. Whether progressive or reactionary in social, economic and environmental grounds, militant particularism entails community building. Once we know how particular communities and urban struggles are produced, we should disclose how connected they are with more universal phenomena such as the circulation of capital, environmental impacts and class inequality (Harvey 2001, pp. 192–196; see also Smith 2001, pp. 41–46 for a critique of a class-based universalism).

Attempts to bridge political economy and contentious politics approaches have not been very explicit in the current literature on urban activism and movements (among the exceptions: Jacobsson 2015b), although empirical research in the field has significantly increased over the last two decades (see, e.g., a wide range of publications about urban squatting: Anders and Sedlmaier 2017; Cattaneo and Martínez 2014; Martínez 2018b; Mudu and Chattopadhyay 2017). A recent analysis of housing movements in line with expanded political economy and intersectional concerns resembles very much the framework we have presented above (Madden and Marcuse 2016). Regarding the links between urban space and protest, there is also a considerable academic production (Carlsson 2001; Mitchell 2003; Nicholls et al. 2013; Routledge 2017; Shephard and Smithsimon 2011). In particular, some have called attention for an explanation of 'multiple spatialities' such as territory, proximity, boundaries, networks, mobility-flows, spatial configurations and patterns that may intersect with social movements (Jessop et al. 2008; Tilly 2003, p. 221). We can skip the risk of spatial determinism (i.e. the assumption that spatialities 'shape' movements, without identifying any clear association of those spatialities with political-economic opportunities and constraints) by accounting for the 'spatial technologies of power' (Miller 2013, p. 290) and 'spatial infrastructures' such as barricades, protest camps (Feigenbaum et al. 2013) and the occupation of vacant property, for instance, when they play relevant roles in urban activism and movements. Despite many social movements may be studied based on their significant 'spatial dimensions' (as well as their historical ones such as duration, acceleration, path-dependence, stages of development, turning points and so on), the analysis of urban movements and activism focuses straightforwardly on their 'spatial claims' (Tilly 2000, p. 137).

Regime change, at the national scale, is also one of the key contextual factors to look at. This seems counter-intuitive given the main drivers for urban movements and activism come from the local scope when they start mobilising. Nonetheless, there is sufficient evidence that urban movements, for example, 'flourish in the phase of collapse of authoritarian regimes and prior to the subsequent formation of political parties' (Pickvance 1995, p. 206) as it was the case in Spain, Portugal, Brazil, Hong Kong, Russia, Poland and Hungary, although 'it is dangerous to generalise about capitalist societies on the basis of what happens in regime transitions' (Pickvance 1995, p. 206). Political opportunity structures are crucial in the analysis of urban movements but other contexts such as the type of welfare regime, the presence of corporatism, the degree of state centralisation and the economic situation and capitalist cycles of growth and stagnation (Pickvance 1995, pp. 212–214) should not be neglected. In sum, the structural emphasis of both the political process and the political economy approaches requires attention to how contextual conditions are articulated with the processes of political agency-subjectivity constitution, creativity-performance and discursive framing by activists themselves (Rutland 2013, p. 997; Marcuse 2009; Smith 2001, pp. 41–46).

Urban Activism Beyond the West

In non-democratic countries, activism that targets state authorities is always being perceived as a threat to the ruling regime. Hence, it is not surprising that collective actions which are not tolerated by the state would in fact face the risk of being repressed. Yet, in non-democratic countries, repression may not always be in the direct form of coercion (e.g. arrest and prosecution). The threat of repression and the use of violence, either directly by the state apparatuses or indirectly by its undercover agents (violence by proxy), are equally effective in suppressing activism (Osa and Schock 2007). However, repression does not always work the way the state authorities expect. Rather than cracking down on activism, repression may instead provoke higher level of contestation. For instance, Almeida (2003) employs examples of activism in El Salvador in the 1960s and 1970s and illustrates how state repression may be able to suppress the action of challenging organisations at the beginning but sustained repression would radicalise activism and provokes even bigger actions. In India and East Timor, escalated state repression triggered backfire and brought about transformative events (Hess and Martin 2006).

Hence, despite the hurdles in organising activism, either because of the personal safety of the organisers being threatened or because it is difficult to mobilise resources for their actions, activism in non-democratic countries does emerge and some of its expressions even have been successful in breeding large-scale transformations. In fact, political opportunity structure that is conducive to activism (access to state institutions, elites' cleavages, allies between movements and powerful parties, social recognition, limited repression, etc.) is still available in non-democratic and authoritarian states.

In the first place, repression of non-democratic regimes is not uniform and cannot always be kept at a high level. As the legitimacy of the state is still an important pillar in sustaining the ruling regime, the capacity (or the promise) to sustain economic growth is often one important instrument in maintaining legitimacy. Hence, when the legitimacy of the state is being challenged by activists, particularly when the country faces economic crisis, repressive measures may be tightened up in order to suppress activism. However, it is also not uncommon for the repressive state to relax social controls which are perceived as a more effective means to ease pressure on the state, particularly when the state launches promises to boost the economy (Osa and Schock 2007).

Secondly, the ruling regime in non-democratic states has cracks. Even in authoritarian states or dictatorship, competition among the ruling factions would create division. Elite division may trigger elite defection which is conducive to the creation of opportunities for challenging organisations (McAdam et al. 2001). Material or moral support for challenging organisations may also be rendered by organisations outside the state institutions or from international organisations. In this respect, pre-existing networked organisations are of eminent significance. These networked organisations may survive from a previous cycle of repression relaxation, as in the case of El Salvador when education and labour organisations were able to preserve their organisational functions (Almeida 2003) or in the case of Poland in the socialist era in which the Catholic Church and the trade unions survived under the repressive regime for respectively cultural and political contingency reasons (Osa 2003).

As Pickvance (1999) argues, 'the fact that citizen organizations of non-political character are permitted allows political activists to operate under non-political labels and to exploit the ambiguity about what is tolerated. And beyond this, independent political activity can exist which can be the basis of social movements' (pp. 358–359). Like the case in Poland, 'non-political' networks help to lay the ground for future social movements as the cost of mobilising would be greatly reduced (Osa 2003). Martin et al. (2001) also illustrate how low-profile non-violent actions can lead to big transformative events in a repressive regime. Like the downfall of Suharto in Indonesia, the sustained actions of a large number of small-scale resistance were able to win over social support and led to the shift of power in the ruling regime which favoured the opposition. As a consequence, small, localised activism on everyday routines may pave the ground for organised actions, movements, political parties and even regime change in the future.

Downfall of the repressive regimes and subsequent migration to a liberal-democratic political system would often generate an expectation of an increased level of activism. On the one hand, discontinuation of the non-democratic regime should lead to the relaxation of repression on activism and on the other hand the incoming democratic regime, which impels an increased level of political participation to enhance its legitimacy, would make articulation of citizen's demand more accommodating. Grievances which have been accumulated in the old regime as well as those that are generated from the economic turmoil that is often associated with political transitions would further push up the level of activism.

However, the experience in the former Eastern Bloc shows contrasting outcomes. A study on the protests in four Eastern European countries (Hungary, East Germany, Slovakia and Poland) over the 1990s (Ekiert and Kubik 1998) indicates that these four countries did not show higher level of protests than their Western European counterparts. Instead, two of the countries even showed a substantially lower level of political mobilisation. In addition, incidents of protest also bear little relation to the economic performance of these new democracies. Countries that experienced a marked economic decline and social dissatisfaction (like Hungary) in the early 1990s had a low level of protest whereas countries that recovered the quickest from the 'transitory recession' (such as Poland) experienced the highest incidence of protests.

Hence, as Ekiert and Kubik (1998) argue, such contrasting outcomes reflect the complex interaction between protests and the changing social and political environment in the democratic consolidation process of transitional economies. There is still no single theory on social movement that is capable of explaining the variation in the protest level but instead a combination of 'resources in a broad sense [which include] traditions, symbols, and discourses alongside material and organizational elements… with the concept of institutional opportunities, which are produced by emerging organizational patterns of the new polity' (p. 581).

Liberalisation of the political institutions also has long-term impacts on how activism is organised. As social and urban movement in liberal democracies is largely sustained by professional social movement organisations, the opening up of the new democratic countries attracted such professional organisations. As a result, activism in the newly democratised countries has undergone a short spell of NGO-isation of movements in the early years of democratisation, largely fuelled by financial and technical support from the West. However, such impetus has not been able to sustain. Dependency on international donors has led to the fragmentation of NGOs in the new democratic countries. New but small organisations mushroomed to compete for the scarce funds. This drives NGOs to align their activities with the interests of international funders away from the concerns of local communities (Jacobsson and Saxonberg 2013). As a consequence, this contributed to weaken the support of the professional movement organisations from the local communities. Detachment between NGO and local communities is further exacerbated by the legacy of the socialist regime on form of general distrust towards political and formal organisations (Jacobsson and Saxonberg 2013).

At the same time, the introduction of neoliberal urban and housing policies into the transitional economies has led to increased dissatisfaction with the reformed regimes. The privatisation of social housing, the gentrification processes, the deterioration of the private housing stock coupled with notorious housing shortages, etc., have widened the gap between the wealthy and the have-nots and generated frustration among ordinary citizens. With organised NGOs alienated from local issues and local communities, this leads to the decline of previous social movements while triggering new forms of small-scale grass-roots activism which are funded locally and target at local problems and needs (Jacobsson 2015a).

Structure of This Volume

This edited volume aims to advance our understanding of urban activism beyond the established theorisation on social movements that has been dominated by thesis of political opportunity structure based on the experience from the global north. In so doing, we collect theoretical and empirical chapters that cover a diversity of urban actions across a broad range of countries in both the east and west hemispheres as well as cities in the global north and global south. One important thread of this volume is to focus on non-institutionalised urban actions that have the potential to bring about structural transformations of the urban system. These forms of urban activism are overlooked by mainstream literature on social movements which tend to concentrate on large-scale processes led by international networks. Also included are actions in authoritarian regimes that are too sensitive to call themselves 'movement'. In fact, such actions are symbolically or materially embedded in the political process but are not associated explicitly with established political parties or state institutions.

This volume is composed of four sections. Section one is an introductory section. In chapter one, we introduce a definition of urban activism in connection to the theoretical approaches in the literature of social movement and urban social movement, in addition to an overview of the topics include in this volume. In Chapter 2, Miguel Angel Martínez López offers a classification of urban activism and movements by highlighting different empirical manifestations and basic theoretical concerns.

Section two pays attention to new and emerging forms of urban activism in the global north and south. This section presents issues that emerge (or re-emerge) at places where few may expect such issues would take root. For instance, the production and circulation of *ignorance* in an economically advanced country (Scotland and the UK) in which state intervention in housing has a significant history. Likewise, struggles for the right to produce food in Athens (Greece), not for leisure but as means of survival and autonomous self-management, represent a sharp contradiction in a mature urban setting at a critical economic stage and political turmoil. A movement for educational rights in China is also a surprising development in a country in which inequality in education is generally being tolerated as the norm although a group of parents dared to launch activists challenge against a tightly controlled authoritarian regime.

Thomas Slater in Chapter 3 explores the ways the Living Rent campaign in Scotland challenged the production and circulation of ignorance in respect of housing affordability and rent control. Slater uses the term of intentional ignorance production, or *agnotology*, to designate the knowledge people could have known or should know but in fact they do not, and traces how such ignorance is intentionally produced by the powerful institutions to retain their vested interests. In particular, the Living Rent campaign confronted the claims that rent controls threaten the quality, supply and efficiency of the housing sector by proposing alternative arguments and reasoning. The chapter demonstrates that the future achievements of housing activism in the UK depend on their ability to strive for policy changes against the backcloth of the production and intensification of ignorance by powerful institutions.

Inés Morales-Bernardos in Chapter 4 examines the urban food activism in Athens in the context of the new neoliberal wave in Europe in general and the slashing of social protection in particular. Challenging the resulting crisis of social reproduction and the rise of far-right activism, the activists organised the reproduction of food in a non-state-centric, autonomous and invisible manner in the urban peripheries. Efforts were made towards de-commodification, the direct and collective organisation of social reproduction, the caregiving and the food provisioning. These forms of activism were to meet everyday needs and thus contributed with new repertoires to the 'politics of feminine'.

Anqi Liu in Chapter 5 looks at the Equal Rights to Education Movement in China as a cultural process. Initially being involved in the movement driven by inflicted personal experience, the participants grew a sense of collective 'we' and solidarity during the movement, which contributed to the change of the goal from addressing personal grievances to striving for education justice. Meanwhile, the movement also provided a venue for the participants to acquire skills in self-organisation, deliberation, and solving internal and external disputes. The Equal Rights to Education Movement is thus considered as a cultural process in which the civic awareness, rights consciousness and autonomy of the Chinese citizens were developed.

Section three focuses on the strategies and tactics of urban activists. Particular attention is paid to activists' networks with the state and other social actors as well as with allies and members. The four chapters in this section offer accounts on how movement organisations confront the harsh social and political environment to push their advocacies forward.

The cases in Hong Kong and Italy both highlight how movements overcome their internal divisions in order to face powerful adversaries.. The cases from Indonesia and Spain compare different organisations on how they strategise their actions and goals. They illustrate how strategies and outcomes are articulated with contingent environmental factors.

Hongze Tan and Miguel Angel Martínez López in Chapter 6 engage in a sociopolitical analysis of cycling advocacy in Hong Kong. In contrast to other global cities in which public dissatisfaction with the 'public transit & private car' transport system contributed to the emergence of civic cyclists who successfully made cycling an increasingly visible issue in the public arena, cycling activists in Hong Kong have yet to struggle for the positioning of urban cycling in Hong Kong, whether it is a recreational leisure activity or an alternative mode of transportation. While cycling activists elsewhere are able to force their governments in responding to their demands, the internal divide among cycling activists in Hong Kong has largely constrained the further development of cycling activism as well as limited the outcome of their advocacy.

Sonia Roitman in Chapter 7 investigates the ways policy decentralisation and community engagement have changed the forms of urban activism in Indonesia. She compares the similarities and differences between two types of activism, one that is initiated by 'deprived' citizens who live in informal settlements demanding poverty alleviation and against social exclusion (called *Kalijawi*) and the other is organised by a diverse group of 'discontented' citizens protesting against the commodification of the city (called *Warga Berdaya*). Despite differences in their origins and strategies, they both advocate for a more just city and the realisation of an equal access to the 'right to the city'.

Robert González in Chapter 8 compares two forms of noteworthy urban movements emerged in Spain as a response to the neoliberal urban-renewal regimes: the pro-housing movement and the squatters' movement. While they both display confluences, yet they should be considered as two different movements. Not only because the former had a more formal organisation with a more integrated identity while the latter was more diversely organised with a more diffuse countercultural identity. It is also because the pro-housing movement focused on specific goals of housing policies while the squatters had broader objectives connected with radical and anti-capitalist traditions. The comparison deepens our understandings of the relations of different stages, political and economic conditions, and outcomes of urban movements.

Gianni Piazza and Federica Frazzetta in Chapter 9 explore the potential of the squatters' movement in Italy to extend their action beyond their locality and reach to the regional, national and global scales. In particular, the authors look at the ways the squatted Social Centres activists, who were labelled as violent by the media and the authorities, participated in two LULU (Locally Unwanted Land Use) movements and performed as central actors bringing generational resources, political-organisational experiences and repertoires of action to the movements. It was found that while the Social Centres activists favoured the cross-issues and cross-territorial scale shift, they still tended to maintain the unity of the movement despite the large differences with other groups.

Section four examines the relationship between urban activism and citizenship, in which the key concern lies in Lefebvre's promulgation of the right to the city. The three chapters in this section explore how the claims on the right to the city can be advocated. The case in Bangkok illustrates how a seemingly technical one-off flood management decision triggered a long-term struggle to claim back the right to the city whereas, on the contrary, the case of Korea takes a long-term perspective in inspecting how activism on the right to housing changes along the development of civil society. On the other hand, the examined case in Buenos Aires addresses the right to the city from the perspective of the negotiations between autonomous movements and mainstream politics.

Danny Marks in Chapter 10 focuses on one particular form of urban activism in the global south: the activism during a 'natural disaster' and the effects of the activism on claims to the right to the city. After the peri-urban fringes of Bangkok were heavily flooded while the inner city was protected by the national government, the floods exposed the vast inequalities that exist in the capital city of Thailand. The unequal governance of the floods led to protests among residents in the flooded areas who asserted that 'we are quality citizens of Bangkok too'. A thorough examination of such activism demonstrates the strong linkage between environmental justice and the right to the city.

Seon Young Lee in Chapter 11 offers a general review of the housing rights activism in South Korea over the last three decades. Three waves of activism were outlined. The first wave was regarded as a direct response to demolition, eviction and financial losses driven by the large redevelopment projects whereas the second wave, termed as owner-occupiers' activism, displayed more complicated conflicts among different social groups. The third wave went even further to demand an alternative urban redevelopment system and to change the conventional concepts of housing and urban development. The trajectory of the housing rights activism suggests that civil society building over time challenged the dominant power relations in a (post-) developmental state.

Ibán Díaz-Parra in Chapter 12 critically examines the extent to which a Squatters and Tenants' Movement (MOI) organisation in Buenos Aires (Argentina) reflects the Lefebvrian concept of the right to the city. The operation of the MOI cooperatives suggests that while the activists believed in the principle of 'self-management, mutual aid, and collective property', they still had to negotiate with the state and comply with the latter's socio-spatial interventions. More importantly, the author probes the ideological dimension of the struggle for urban centrality and addresses a key political question on the possibility or impossibility of transforming socio-spatial orders.

References

Almeida, P. D. (2003). Opportunity Organizations and Threat-Induced Contention: Protest Waves in Authoritarian Settings. *American Journal of Sociology, 109*(2), 345–400.

Anders, F., & Sedlmaier, A. (2017). *Public Goods Versus Economic Interests. Global Perspectives on the History of Squatting*. New York: Routledge.

Andretta, M., Piazza, G., & Subirats, A. (2015). Urban Dynamics and Social Movements. In D. Della Porta & M. Diani (Eds.), *The Oxford Handbook of Social Movements* (pp. 200–218). Oxford: Oxford University.

Carlsson, C. (Ed.). (2001). *Critical Mass. Bicycling's Defiant Celebration*. Oakland: AK Press.

Castells, M. (1972 [1974]). *La cuestión urbana*. Madrid: Siglo XXI.

Castells, M. (1983). *The City and the Grassroots. A Cross-Cultural Theory of Urban Social Movements*. Berkeley: University of California.

Cattaneo, C., & Martínez, M. (Eds.). (2014). *The Squatters' Movement in Europe: Commons and Autonomy as Alternatives to Capitalism*. London: Pluto Press.

Della Porta, D. (2015). *Social Movements in Times of Austerity*. Cambridge: Polity.

Dikec, M., & Swyngedouw, E. (2017). Theorizing the Politicizing City. *International Journal of Urban and Regional Research, 41*(1), 1–18.

Ekiert, G., & Kubik, J. (1998). Contentious Politics in New Democracies: East Germany, Hungary, Poland, and Slovakia 1989–93. *World Politics, 50*(4), 547–581.

Fainstein, S. S., & Hirst, C. (1995). Urban Social Movements. In D. Judge, G. Stoker, & H. Wolman (Eds.), *Theories of Urban Politics* (pp. 181–204). London: Sage.

Feigenbaum, A., Frenzel, F., & McCurdy, P. (2013). *Protest Camps*. London: Zed.

Goodwin, J., & Jasper, J. (Eds.). (2004). *Rethinking Social Movements: People, Passions, and Power*. Lanham, MD: Roman and Littlefield.

Hamel, P. (2000). The Fragmentation of Social Movements and Social Justice. Beyond the Traditional Forms of Localism. In P. Hamel, H. Lustiger-Thaler, & M. Mayer (Eds.), *Urban Movements in a Globalising World* (pp. 158–176). London: Routledge.

Hamel, P., Lustiger-Thaler, H., & Mayer, M. (2000). Introduction. Urban Social Movements—Local Thematics, Global Spaces. In P. Hamel, H. Lustiger-Thaler, & M. Mayer (Eds.), *Urban Movements in a Globalising World* (pp. 1–21). London: Routledge.

Harvey, D. (1973). *Social Justice and the City*. Athens, GA: University of Georgia.

Harvey, D. (1982). *The Limits of Capital*. Oxford: Blackwell.

Harvey, D. (1996). *Justice, Nature & the Geography of Difference*. Cambridge: Blackwell.

Harvey, D. (2001). *Spaces of Capital. Towards a Critical Geography*. New York: Routledge.

Harvey, D. (2008). The Right to the City. *New Left Review, 53*, 23–40.

Heatland, G., & Goodwin, J. (2013). The Strange Disappearance of Capitalism from Social Movement Studies. In C. Barker, L. Cox, J. Krinsky, & A. Gunvald (Eds.), *Marxism and Social Movements* (pp. 83–102). Brill: Leiden.

Hess, D., & Martin, B. (2006). Repression, Backfire, and the Theory of Transformative Events. *Mobilization, 11*, 249–267.

Jacobsson, K. (2015a). Introduction: The Development of Urban Movements in Central and Eastern Europe. In K. Jacobsson (Ed.), *Urban Grassroots Movements in Central and Eastern Europe*. Farnham: Ashgate.

Jacobsson, K. (Ed.). (2015b). *Urban Grassroots Movements in Central and Eastern Europe*. Farnham: Ashgate.

Jacobsson, K., & Saxonberg, S. (2013). Introduction: The Development of Social Movement in Central and Eastern Europe. In K. Jacobsson & S. Saxonberg (Eds.), *Beyond NGO-ization: The Development of Social Movements in Central and Eastern Europe*. London and New York: Routledge.

Jessop, B., Brenner, N., & Jones, M. (2008). Theorizing Sociospatial Relations. *Environment and Planning D: Society and Space, 26*, 389–401.

Lowe, Stuart. (1986). *Urban Social Movements. The City After Castells*. London: Macmillan.

Madden, D., & Marcuse, P. (2016). *In Defense of Housing. The Politics of Crisis*. London: Verso.

Marcuse, P. (2002). Depoliticizing Globalization: From Neo-Marxism to the Network Society of Manuel Castells. In J. Eade & C. Merle (Eds.), *Understanding the City. Contemporary and Future Perspectives* (pp. 130–158). Oxford: Blackwell.

Marcuse, P. (2009). From Critical Urban Theory to the Right to the City. *City, 13*(2–3), 185–197.

Martin, B., Varney, W., & Vickers, A. (2001). Political Jiu-Jitsu Against Indonesian Repression: Studying Lower-Profile Nonviolent Resistance. *Pacifica Review: Peace, Security & Global Change, 13*(2), 143–156.

Martin, D. G., & Miller, B. (2003). Space and Contentitous Politics. *Mobilization: An International Journal, 8*(2), 143–156.

Martínez, M. (2018a). Bitter Wins or a Long-Distance Race? Social and Political Outcomes of the Spanish Housing Movement. *Housing Studies*, https://www.tandfonline.com/doi/full/10.1080/02673037.2018.1447094.

Martínez, M. (Ed.). (2018b). *The Urban Politics of Squatters' Movements.* New York: Palgrave Macmillan.

Mayer, M. (1993). The Career of Urban Social Movements in German Cities. In R. Fisher & J. Kling (Eds.), *Mobilizing the Community: Local Politics in a Global Era* (pp. 149–170). Newbury Park: Sage.

Mayer, M. (2006). Manuel Castells' the City and the Grassroots. *International Journal of Urban and Regional Research, 30*(1), 202–206.

Mayer, M. (2009). The 'Right to the City' in Urban Social Movements. *City, 13*(2–3), 367–374.

Mayer, M. (2016). Neoliberal Urbanism and Uprisings Across Europe. In M. Mayer, C. Thörn, & H. Thörn (Eds.), *Urban Uprisings: Challenging the Neoliberal City in Europe* (pp. 57–92). London: Palgrave Macmillan-Springer.

Mayer, M., & Boudreau, J.-A. (2012). Social Movements in Urban Politics: Trends in Research and Practice. In P. John, K. Mossberger, & S. E. Clarke (Eds.), *The Oxford Handbook of Urban Politics* (pp. 273–291). London: Oxford University Press.

McAdam, D., Tarrow, S. G., & Tilly, C. (2001). *Dynamics of Contention.* Cambridge, New York: Cambridge University Press.

Merrifield, A. (2014). *The New Urban Question.* London: Pluto.

Miller, B. (2013). Conclusion: Spatialities of Mobilization: Building and Breaking. In W. Nicholls, B. Miller, & J. Beaumont (Eds.), *Spaces of Contention. Spatialities and Social Movements* (pp. 285–298). Farnham: Ashgate.

Mitchell, D. (2003). *The Right to the City. Social Justice and the Fight for Public Space*. Guilford: New York.

Mudu, P., & Chattopadhyay, S. (Eds.). (2017). *Migration, Squatting and Radical Autonomy.* Abingdon: Routledge.

Nicholls, W., Miller, B., & Beaumont, J. (2013). Introduction: Conceptualizing the Spatialities of Social Movements. In W. Nicholls, B. Miller, & J. Beaumont (Eds.), *Spaces of Contention. Spatialities and Social Movements* (pp. 1–23). Farnham: Ashgate.

Osa, M. (2003). Networks in Opposition: Linking Organizations Through Activists in the Polish People's Republic. In M. Diani & D. McAdam (Eds.), *Social Movements and Networks: Relational Approaches to Collective Action*. Oxford: Oxford University Press.

Osa, M., & Schock, K. (2007). A Long, Hard Slog: Political Opportunities, Social Networks and the Mobilization of Dissent in Non-Democracies. *Research in Social Movements, Conflicts and Change, 27,* 123–153.

Pickvance, C. G. (1975). On the Study of Urban Social Movements. *The Sociological Review, 23*(1), 29–49.

Pickvance, C. G. (1985). The Rise and Fall of Urban Movements and the Role of Comparative Analysis. *Environment and Planning D: Society and Space, 3,* 31–53.

Pickvance, C. G. (1995). Where Have Urban Movements Gone? In C. Hadjimichalis & D. Sadler (Eds.), *Europe at the Margins: New Mosaics of Inequality* (pp. 197–217). London: Wiley & Sons.

Pickvance, C. G. (1999). Democratisation and the Decline of Social Movements: The Effects of Regime Change on Collective Action in Eastern Europe, Southern Europe and Latin America. *Sociology, 33*(2), 353–372.

Pruijt, H. (2007). Urban movements. In G. Ritzer (Ed.), *Blackwell Encyclopedia of Sociology*. Malden: Blackwell.

Rossi, U. (2017). *Cities in Global Capitalism*. Cambridge: Polity.

Roth, R. (2000). New Social Movements, Poor People's Movement and the Struggle for Social Citizenship. In P. Hamel, H. Lustiger-Thaler, & M. Mayer (Eds.), *Urban Movements in a Globalising World*. London and New York: Routledge.

Routledge, P. (2017). *Space Invaders. Radical Geography of Protest*. London: Pluto.

Rutland, T. (2013). Activists in the Making: Urban Movements, Political Processes and the Creation of Political Subjects. *International Journal of Urban and Regional Research, 37*(3), 989–1011.

Saunders, P. (1983). *Urban Politics. A Sociological Interpretation*. London: Hutchinson & Co.

Scott, J. C. (1990). *Domination and the Arts of Resistance: Hidden Transcripts*. New Haven: Yale University.

Shephard, B., & Smithsimon, G. (2011). *The Beach Beneath the Streets. Contesting New York City's Public Spaces*. Albany: Excelsior-State University of New York.

Smith, M. P. (2001). *Transnational Urbanism. Locating Globalization*. Malden: Blackwell.

Tilly, C. (2000). Spaces of Contention. *Mobilization: An International Journal, 5*(2), 135–159.

Tilly, C. (2003). Contention Over Space and Place. *Mobilization: An International Journal, 8*(2), 221–225.

Tilly, C., & Tarrow, S. (2007). *Contentious Politics*. New York: Oxford University Press.

Tyler, I., & Slater, T. (2018). Rethinking the Sociology of Stigma. *The Sociological Review Monographs, 66*(4), 721–743.

CHAPTER 2

Framing Urban Movements, Contesting Global Capitalism and Liberal Democracy

Miguel Angel Martínez López

Introduction

In this chapter, I define the notion of 'urban movements' according to a theoretical framework in which the contestations of global capitalism and liberal democracy are central concerns. These concerns help to distinguish types and cases of urban movements. I focus here on the actual and potential contributions made by different forms of collective action to improve our cities. However, all movements' contradictions and limitations and less progressive or even conservative forms of urban activism deserve to be carefully investigated as well. Rather than a detailed examination of the academic literature (Andretta et al. 2015; Mayer and Boudreau 2012; Martí and Bonet 2008; Nicholls et al. 2013; Pickvance 1995; Pruijt 2007), I offer a general approach to the topic based on my past research and a few contemporary examples.

In my view, democratic societies ought to deal in a sensible and sensitive manner with all the voices and bodies of their members as

M. A. Martínez López (✉)
Uppsala University, Uppsala, Sweden
e-mail: miguel.martinez@ibf.uu.se

N. M. Yip et al. (eds.),
Contested Cities and Urban Activism, The Contemporary City,
https://doi.org/10.1007/978-981-13-1730-9_2

inhabitants (some not legally recognised as citizens), especially those who are ignored and marginalised from the crucial decision-making processes. However, the prevailing economic and political elites have set up many mechanisms to prevent a meaningful attention to those at the bottom of power structures based on class, gender, ethnicity, age, abilities, knowledge and other social conditions. Liberal and representative democracies are far from meeting the needs, rights and aspirations of the worst off. Capitalism is also an endless source of authoritarian rule, from the workplace to the stock markets. In addition, an increasing concentration of wealth on a global scale erodes any horizon of more equal and sustainable societies.

There are multiple social movements in general and urban movements in particular that question this state of affairs all over the world, not only in formally liberal democracies, and not only related to the urban fabric. When formed by social groups who are powerless within the established institutions, these movements activate their own capacities, knowledge and social alliances, mostly apart from state institutions and by confronting them. As a consequence, to raise the political leverage of inhabitants usually means increasing conflicts as well (Martínez 2011). A comprehensive research agenda should also assess the outcomes of urban struggles in terms of the specific changes, if any, achieved by movements or due to other circumstances. This entails a critical inquiry about the contradictory processes in which urban movements are centrally involved, and an examination of their development over time according to the significant contexts that constrain them.

In the next section, I briefly define the meaning of urban movements. Secondly, I distinguish them in four categories. Both the definition and the classification will be illustrated with some references to cases from Southern Europe and Latin America. Finally, I will elicit some implications of this approach in order to understand the role played by social movements in urban politics.

Activism, Movements, Contexts

Urban movements may be defined as sustained collective actions of claim-making in the production, governance and change of cities, according to specific societal contexts (i.e. the social articulation of political, economic and cultural structures). Citizens usually come together as a response to specific grievances. They self-organise and perform various types of protest in order to achieve their goals. Sometimes movement participants may use institutional channels to express their concerns, but authorities and other opponents do not always react in a satisfactory manner. Therefore, many activists resort to direct actions. Most urban movements choose to go non-institutional from the very beginning by assuming that this is the most available and effective way to challenge the dominant status quo of cities in terms of cultural values, power structures and economic inequalities.

Urban gardeners and protesters against urban renewal and social displacement, for instance, may be among current urban movements with a progressive agenda, although their political motivations and unintended effects must be assessed on a case by case basis. On the more conservative side, there are groups of residents from middle and upper classes who intend to exclude migrants and certain public facilities from their privileged neighbourhoods because, according to their rationale, they can increase criminality and jeopardise property values. These exclusionary campaigns can take place in some community gardens and underprivileged urban areas as well. There are many more disparate examples of urban movements: advocates of urban cycling, rent strikes, organised slum dwellers, reclaim the streets mobilisations, environmental justice against polluting infrastructures, neighbourhood crime watch patrols, etc. They all contribute to shaping the urban landscape for good or for bad, depending on their intentions and actual influence.

In liberal democracies, and sometimes also in more authoritarian regimes, citizens are expected to channel their demands through voting, formal interest groups, political parties and the municipal bureaucratic administration. However, these means might not be efficient and may engender distrust among the citizenry should they systematically end up in reproducing prevailing elitist structures. Therefore, urban activism makes claims that are not fully satisfied by the local authorities. When these claims escalate to coordination between different social groups, organisations, supporters, sympathisers and even allies within state institutions, urban movements can emerge, persist and tackle the power structures of cities. This means that movements are made of activists and urban activism, but not all forms of activism turn into a movement.

To illustrate this first distinction, I suggest looking at the rising cry about the improvement of conditions for the circulation of bicycles in cities. A call to install just one bicycle rack in one building amounts to a very low form of contentious activism. This kind of 'urban claim' often addresses no more than individual representatives at the town hall or the managers of the building. If the claims are about the extension of cycling facilities all over the city and advocacy organisations take the lead, their activism may involve a broader span of time, space and political challenge to the local authorities. The breadth and numbers of the claims turn them into activist campaigns. These activists may engage in hearings, forums, objections to master plans and participatory processes within the institutional realm. The more they surpass a moderate repertoire of protest, the more they mobilise supporters through media, demonstrations, direct actions and road blockades on a regular basis, and the more they are prone to shape a singular 'urban movement'. An illustration of the latter is the Critical Mass. This mobilisation of urban cyclists started in San Francisco in 1992 and became a popular form of protest in hundreds of cities worldwide (Carlsson 2001; Shephard and Smithsimon 2011, pp. 44–49). Critical Mass is labelled by its promoters as a celebration or a playful display of creative dissent. It usually takes place once a month in many cities, sometimes simultaneously. Hence, urban movements can also be diffused transnationally. As such, they are not entirely attached to a given city or local space. Critical Mass also represents an opposition to the dominant motorisation of cities and claims more bicycle-friendly urban environments. Their regular demonstrations, articulated with other formal advocacy groups, certainly contributed to the relative revival of urban cycling in transport policies over the last decades.

In contrast to open claims, activist campaigns and urban movements, it may happen that collective actions of protest are performed through silent, clandestine, informal and hidden ways either when the political conditions are considered oppressive by activists or when they just want to avoid too much public exposure. This type of urban infra-politics (Scott 1990) rejects clear identities and prefers loose networks of self-organisation. For example, painting signs on the roads to favour cycling and tactical daily manoeuvres to obstruct motorised traffic may be part of this diffuse repertoire. Finally, closely related to activism although not identical, 'urban riots' may occur (Mayer et al. 2016). For example, when clashes between cyclists, the police and motorists erupt, based on the frustration that many cyclists (and pedestrians) experience as subaltern vehicles. The violent development of urban conflicts is often related to underlying injustices and suppressed forms of protest, but they are not necessarily a frequent manifestation of social movements.

Table 2.1 Forms of collective action in urban politics

	Duration-coordination	*Contention*	*Institutions*
Urban movements	▲▲▲	▲▲▲	▲
Urban riots	▲	▲▲▲	▲
Urban clandestine activism	▲▲	▲	▲
Urban activism	▲▲	▲▲	▲▲
Political parties	▲▲▲	▲▲▲	▲▲▲

Source Author

According to the above categories, a possible classification of urban collective actions would hinge upon three main criteria: (1) the duration of coordinated activism; (2) the intensity of contentious processes; and (3) the degree of involvement in state institutions. These criteria help distinguish urban movements from other types of collective action in urban politics, especially those more embedded in the institutional realm such as competing political parties, but also from those lasting a shorter time and less coordinated in dense social networks—particular activist claims and campaigns, clandestine activism and riots. However, as the dotted lines suggest, there are also many empirical intersections and reciprocities between all the categories. In fact, specific claims and onetime protests may trigger the engagement of activists who, in turn, may nurture broader mobilisations, campaigns and political identities (Tilly and Tarrow 2007) (Table 2.1).

Housing and squatters' movements provide a striking manifestation of this approach while also expressing the deep societal conflicts and contestations involved in their significant context. With the Global Financial Crisis in 2008, unemployment and homelessness skyrocketed, especially in Southern Europe. Poor migrants and refugees were the most vulnerable groups hit by the nightmare created by financial speculation all over the world and the dismantling of welfare services since the decade of the 1980s. Activists from different nationalities and political backgrounds reacted to this dramatic situation. For example, I observed many minors who participated in a demonstration for the right to housing which I attended in Rome, in 2016. It tells a dramatic story of poverty and violation of basic human rights at the heart of a wealthy country. These children and their parents are among the thousands of people who are homeless in Rome—official figures only mention 3300 by 2015, but there are many more who struggle to get a place in temporary shelters and dream of a decent, affordable and permanent home. In Rome, three main organisations united and launched a campaign called Tsunami Tour in October 2012. Every time they gathered, they took over various empty buildings simultaneously: 10 buildings were squatted on 6 December 2012, 14 buildings more on 6/7 April 2013, four more in October 2013 and another six in April 2014 (Di Feliciantonio 2017). These are just the most visible cases of developed coordination and high contentious stakes. Nonetheless, the occupation of abandoned buildings to house people in extreme need is a current practice, usually in a hidden, silent and less organised manner, without the assistance of political and grassroots groups. Very often, squatters are brutally evicted by riot police. Little children, the elderly, ill and disabled people are usually thrown back in the streets too. When they do not hold documents of full citizenship, their chances to access social housing or overcrowded shelters are very low, in addition to a more likely risk of being arrested, and even deported if other circumstances apply.

For example, in just one operation, about 800 people, mostly asylum seekers from Eritrea, were forcibly expelled from Palazzo Curtatone refugee squat in Rome, on 20 August 2017. More than 500 police officers executed the eviction. Two pregnant women were among the evicted squatters. When the refugees occupied a nearby public square to protest, they were also removed by the police who used water cannons and batons. The Palazzo Curtatone had been squatted since 2013, and no housing alternatives were provided to the refugees.[1] In the coverage of this case by the economic newspapers, journalists usually portrayed squatters as criminals and the occupation as unacceptable damage to the profits of the owners.[2] In particular, the investment property firm Omega from Idea Fimit, a company gaming with institutional and pension funds, bought this building in 2011 for 75 million Euros and left it empty until they could get a better deal after renovation or by selling it to a third party. While the figures of around 30,000 home evictions in the city of Rome between 2003 and 2015 (Di Feliciantonio 2017) represent a striking failure of public policies, the

[1] https://enoughisenough14.org/2017/08/24/rome-cops-evict-refugee-squatters-from-palazzo-curtatone-square/.

[2] http://carlofesta.blog.ilsole24ore.com/2017/08/24/lo-sgombero-di-palazzo-curtatone-i-falsi-buonismi-e-le-perdite-milionarie-dei-fondi-pensione/?refresh_ce=1 and http://www.ilsole24ore.com/art/notizie/2017-08-24/immobile-sgomberato-roma-quattro-milioni-euro-perdite-le-casse-pensionati--124536.shtml?uuid=AEBRK3GC.

profits made by real-estate operators, their tax evasions and their professional lobbying of urban regulations remain largely opaque.

Similar stories can be told about Greece, France and Spain where housing movements allied with homeless, immigrants and refugees are not passive spectators anymore. In Spain, for example, 387,966 properties have been evicted between 2011 and 2015.[3] These data include all kind of properties and legal conditions, not only mortgaged residential ones, but it is estimated that more than half of them represent primary residence homes.[4] As a response, the Platform for People Affected by Mortgages (PAH) was established in the metropolitan area of Barcelona in 2009. By May 2016, 236 local PAH groups were active across Spanish cities.

Initially, the organisation's main concern was evictions of people unable to pay off mortgage loans. The PAH was formed to protect individuals and families from foreclosure. Considering the high rates of unemployment that followed the 2008 Global Financial Crisis, the PAH quickly attracted thousands of people threatened with eviction, not only former homeowners who became financially broke. PAH members and supporters evolved into a broader social base including migrants, working- and middle-class people, activists from other movements and even sympathetic politicians. Many women and migrants became extraordinarily empowered by joining this and other allied organisations, especially those arising during the 2011 Indignados or 15M movement. PAH members shared their experiences communally and developed strong capabilities of mutual aid, mobilisation and negotiation. This implies a self-help approach aimed at preventing repossession as well as mitigating the hardships experienced by these impoverished groups.

Interestingly, the demands of the PAH convey an interpretation of the 2008 economic recession that is in sharp contrast to that offered by mainstream media and the dominant elites. PAH activists, guided by thorough research on the matter, pointed to the financialisation of property assets, the construction bubble, abusive and deregulated banking practices, austerity policies, meagre earnings, unavailable public housing and the corrupt practices of major political parties as the main driving forces behind the devastating wave of foreclosures and housing repossessions. At the peak of

[3] General Council of the Judicial Power: http://www.poderjudicial.es/portal/site/cgpj/.

[4] http://www.bde.es/f/webbde/GAP/Secciones/SalaPrensa/NotasInformativas/Briefing_notes/es/notabe300715.pdf.

the crisis, between 2011 and 2015, direct actions aiming to prevent home evictions were framed by the media with a positive outlook for the PAH. The Stop Evictions campaign entailed not only the risk of being removed, beaten and arrested by the police, but also criminal charges for obstructing court injunctions. The outlook of radicalism usually attached to direct actions, even those strictly adherent to nonviolent civil disobedience principles, was largely compensated for by the images of evicted individuals and families who had been offered neither compassion nor policy measures to house them. According to the last check of their web page (October 2017), the PAH were able to prevent more than 2000 cases of home evictions and helped rehouse more than 2500 people, very often in squatted buildings.[5] This political activity, in my opinion, contributed to making cities more liveable for those who participated and benefitted from the struggle, despite not solving the problem at large (Martínez 2018a) (Fig. 2.1).

Fig. 2.1 Demonstration for the right to housing, Barcelona (Spain), 2017 (*Source* Photograph by the Author)

[5]http://afectadosporlahipoteca.com/.

Commons, Rights, Participation and Communities

The Global Financial Crisis marked a turning point that triggered the rise of such movements, but the structural roots of these protests in Southern Europe and elsewhere lay in more long-term processes. Capitalists and neoliberal politicians aim to fuel real-estate speculation in every city corner which can return a profit for investors, regardless of the side effects such as rising prices, inflation, housing unaffordability and displacement of those who cannot stand up after the shock. Cuts in welfare benefits and services, and especially in social housing, are an additional burden. Casual employment and precarious working conditions are nowadays more widespread. Both national and European migration and asylum policies increase vulnerability, civic death and labour exploitation of newcomers to wealthier countries who have left their own ruined lands where Western-Global-North corporations and governments hold vested interests. Hence, the origins of local movements are not only at the local scale, but they are also defined by a broader national and international scope of political and economic processes.

As the examples presented above make clear, sometimes urban movements do not question just local authorities but also the main economic actors, structures and regulations that cause a miserable life for millions of urban dwellers. These observations allow me to introduce a more useful classification of urban activism and movements.

First of all, some urban movements are focused on what can be designated as 'urban commons', in line with previous categories such as 'urban trade-unionism' or struggles for 'collective consumption'. These include all protests against housing shortages and rent increases, tenants' unions and the squatting of vacant buildings and land. They all point to the redistribution of public resources which is in turn an essential dimension of capitalism in order to reproduce the labour force. But capitalists are not so interested in a fair and extensive redistribution of wealth, especially if they can make more profits out of the privatisation of welfare services. Movements defending the urban commons set social barriers to the overwhelming commodification of life. These mobilisations tend to involve forms of self-organisation that push public policies beyond their bureaucratic ethos, not only in the field of housing but also in the provision, access, maintenance and quality of local health care and education facilities, water and electricity supply, sewerage, public transport and food.

Bear in mind, for example, informal and squatted settlements, favelas, shanty towns or slums that can be found in most megacities of the Global

South and also in the urban fringes of the Global North. Just think on the following figure: one billion people. More precisely, more than 900 million of people are considered slum dwellers in developing countries according to the United Nations (UN-Habitat 2016). This means that one in eight people live in slums nowadays, and the absolute numbers are growing, despite a relative decrease from 39 to 30% between 2000 and 2014. Insecure tenure, lack of access to drinking water and sanitation facilities and poor housing structures are some of the complex features faced by one-eighth of the world population. Wealth redistribution and improvement of the urban commons is an urgent cry in all these settlements. Inhabitants' organisations and self-empowering initiatives are especially salient there, in addition to the international aid. There are activist networks with different degrees of institutionalisation such as Slum/Shack Dwellers International, Asian Coalition for Housing Rights, Habitat International Coalition and the International Alliance of Inhabitants that offer support through advocacy, documenting experiences and conducting research. To recall a case I am familiar with, the massive operations of rehousing informal settlers in Madrid (Spain) in the 1980s, for example, involved significant experiences of participatory planning, the creation of housing cooperatives and the enthusiastic involvement of activists, professionals and researchers (Castells 1983; Villasante et al. 1989). On another geographical site, autonomous organisations such as the *piqueteros* from different *villas miseria* in Argentina championed class-based mobilisations and obtained subsidies from favourable governments until the recent shift in the central government. Due to their persistent grassroots organisations, some Brazilian favelas have been consolidated as urban neighbourhoods whereas others were not able to resist their brutal eradication due to urban mega-projects such as the Olympics Games held in Rio de Janeiro in 2016.[6]

In order to be alert about 'counter-movements' in the field of urban redistribution of wealth, it is also worth mentioning the occasional tax-revolts campaigns run by well-off residents. Similarly, privileged inhabitants can efficiently lobby and mobilise in a larger scale if necessary in order to attract desired public facilities or improve the existing ones in their residential area. This activism may thus increase the territorial unbalance of state investments at the city or regional scale regardless of any thorough analysis of needs and inequality gaps.

[6]http://www.rioonwatch.org/?p=37894.

A second group of movements, often closely associated and intertwined with the former, are focused on the 'right to the city' (Marcuse 2009; Mayer and Boudreau 2012, p. 280; Mitchell 2003). They incorporate the spatial claims expressed by particularly deprived groups which are not only defined by their socioeconomic class condition. In the interpretation I prefer to emphasise here, the right to the city accounts for the appropriation of urban spaces (streets, plazas, green areas and especially the city centre) by those more in need, excluded, dispossessed, exploited and alienated by capitalist forces. The angle taken by the 'right to the city' illuminates the spatial practices, needs and demands of groups such as women, non-white and minority ethnic groups, migrants targeted by police raids, the youth, the elderly, disabled people, homeless people, prostitutes, indigenous populations, diverse gender and sexual identities, street vendors, urban outcasts and others.

Struggles against environmental hazards and risks affecting vulnerable groups represent the intersection between the right to the city and environmental justice. More pedestrian and bicycle-friendly cities as promoted by movements such as reclaim the streets and the Critical Mass put the rights of non-car owners at front. Another example is a mural drawn in the walls of San Lorenzo neighbourhood in Rome (Italy) aims to remember the 105 women murdered in Italy in 2012. Although 63% of these murders occurred inside domestic spaces,[7] the female names and icons are also a sharp reminder of how unsafe many public spaces are for women and Lesbian, Gay, Bisexual, Trans-sexual, Intersexual, Queer (LGBTIQ) people in most cities of the world, especially at night. Harassment-free spaces are not just one of the many concerns of feminist movements, but also a matter that should concern all males, planners, the police and community organisations. In many countries, official figures of gender violence are not a good indicator because these crimes remain largely under-reported, but there is abundant evidence of 'femicide', for instance, from the tragic well-known case of Ciudad Juárez in México where more than 500 women between 1994 and 2010 were found dead, many after being raped, tortured and mutilated (Sweet and Ortiz 2010) (Fig. 2.2 illustrates how feminist and LGBTIQ claims permeate urban movements such as the squatting of urban land).

[7] http://www.npr.org/2012/11/23/165658673/italian-women-call-for-action-against-femicide.

Fig. 2.2 Concert at the occupied community garden *Ágora Juan Andrés*, Barcelona, 2017 (*Source* Photograph by the Author)

The label 'right to the city' may also be perverted when it is turned into a politically correct language as the 'right to the city for all'. Many municipal and metropolitan authorities adhere to its all-encompassing reference in order to guarantee, for example, the demands of private companies to occupy public spaces for all kind of commercial activities. If they do not enjoy this access for free and according to their business interests, they manifest a grievance of exclusion and marginalisation, as if they were victims of the emancipatory actions of the subaltern groups. Although those commercial demands are rarely channelled via activist campaigns on their own, they can result in tactical manoeuvres where private firms sponsor festivities, rallies, pride parades, charity events, artists' gatherings and exhibitions (and squats), etc.

A third type of urban activism that may also end up in notable urban movements is represented by citizens' participation in urban plans and policies, and their critical voices when they feel excluded. Feminist movements, for instance, have been also involved in these dimensions of urban planning when questioning dominant transport policies, the location of schools, the design of streets and sidewalks, etc. (Ortiz and Gutiérrez 2015). Other mobilised residents, usually with the help of 'advocacy planners' and other experts, may propose alternative spatial plans and tactical urbanism, measures for political and administrative decentralisation and collective assessment of the effects of urban displacement, gentrification, tourism, mega-events and corporate urban development. They can be either cooperative or disruptive parties in urban governance when they dispute urban renewal projects. Anti-neoliberal initiatives to prevent the privatisation of local services and public space, or to establish public-state partnerships guided by the principles of economic redistribution and social well-being, allow activists to participate in decision-making processes. This contested ground of urban politics involves deliberative and regulatory processes that aim at protecting the urban commons and the rights of the many while limiting the market forces.

As far as capitalists try to make profits out of every spatial asset by manipulating legal frameworks and exploiting social capacities, it is not only the urban space within the municipal boundaries which is at stake, but metropolitan regions and spaces elsewhere too (Harvey 2016; Logan and Molotch 1987). Accordingly, social movements attached to specific urban places may also question and interrupt the circulation of capital wherever the spaces are in which it manifests.

Participatory urban governance has become a very moderate and institutionalised programme at best, and very frequently a legitimation of ready-made policies. So-called radical urban movements are not often interested in participatory planning or not invited to take part. However, sometimes non-institutional protests may have some effects in urban policies (Martínez 2011). Take, for example, the demonstrations against the adverse impacts of tourism in the city of Barcelona. Over the last decade, community organisations criticised the expansion of illegal tourist flats, the nuisances produced by the leisure and nightlife businesses catering to tourists, and the displacement of both traditional retail stores and tenants unable to meet rent increases.

Since 2015, the municipal government of Barcelona has launched several policies aiming at inspecting the operations of companies such as Airbnb and halted the opening of new hotels in the central areas. Airbnb has even admitted that around 78% of the apartments hosted on their website in Barcelona are illegal (Nofre et al. 2017). The city has an official population of 1.6 million but has been the destination of 3.5 million visitors in 2000, and eight million in 2014. By means of media campaigns, the new progressive government has also targeted tourists in order to soothe the tensions that this overcrowding touristification generates. Anyhow, the main economic structures that fuel the tourist industry remain largely untouched and resist most attempts to threaten their rising profits.

A final fourth category of urban movements comprises those of a more 'localist' and 'communitarian' nature. In these cases, activists aspire to greater control over their territory regardless of the interests of the whole city or region. Local assemblies and homeowners' organisations are the usual expressions of communitarianism. They can also self-manage parks, local newsletters, radio-tv stations and festive events in the public sphere within the boundaries of a given neighbourhood, district or urban area. Liveability and quality of life are their major concerns.

In some cases, these movements may turn into unfair and exclusionary campaigns against migrants and temporary labourers in the streets, or by refusing to pay municipal taxes. Not In My BackYard (NIMBY) and Locally Unwanted Land Use (LULU) protests oppose infrastructure, buildings and services that can be of general interest for citizens beyond its location. NIMBYs and LULUs tend to be restricted to very specific places, but they can also turn into urban movements when they become articulated with other groups, campaigns and mobilisations. This is the case with the opposition to military facilities and projects with a damaging impact in the environment such as the high-speed train in the north of Italy (Piazza 2011). Activists may also keep watch on police stop-and-frisk practices targeting migrants, ethnic minorities and youth in particular urban areas (Martínez 2017). Heritage preservation, even by resorting to tactical squatting (Pruijt 2007), usually falls under the umbrella of communitarianism.

In other cases, as has happened in many parts of the densely populated Mexico Federal District, squatted settlements have turned into legalised neighbourhoods that received public funds to build their houses. In addition, residents developed strong and autonomous self-organisations. A paradigmatic case is La Polvorilla, a 'colony' that houses around 3000 dwellers. It is affiliated with an umbrella housing cooperative, Acapatzingo. Ten similar settlements were promoted by the political group behind them, the Frente Popular Francisco Villa. This and many other urban organisations scaled up their coordination in the aftermath of the 1985 earthquake as a reaction to a clientelistic and authoritarian state unable to offer help. A similar phenomenon occurred quite recently, in September 2017, when another intense earthquake gave birth to renewed networks of grassroots solidarity. Matters inside the community are managed according to general assemblies and 'brigades' following the colour of buildings. A significant feature of La Polvorilla is the two main gates that prevent the police from coming inside the neighbourhood. Although this resembles upper-class gated communities with their own security guards, the measures taken by the poor residents of La Polvorilla grant them a relative safety from the rampant corruption among the local police and oblige them to improve their own mechanisms to solve internal conflicts (Pineda 2013; Stavrides 2017).

Movements as City Makers

In a recent interview, urban scholar Richard Florida recalled the case of a puzzled student who came to his office and said 'Everything is gentrification now!... I took this class in urban geography and I want to make my city better, but they say everything I want to do is gentrification. A better school is gentrification, empowering artists is gentrification, working to improve the condition of parks is gentrification. What can I do?'[8] Florida is a worldwide academic celebrity. He founded a very lucrative consultancy, Creative Class Group, and offers speeches for a minimum fee of $35,000. For more than a decade, he was advising local politicians with a straight message: invest in urban clusters where creative and innovative professionals, i.e. hipsters and techies, can settle down, freely interact and enjoy cultural life; this will generate more economic prosperity than mega-infrastructures and tax breaks to lure private companies. Many mayors followed the recipe but bought the whole package—facilities and expensive infrastructures for the big tycoons, and support to creative people in the arts scene and start-up technologies. These were the apparent driving forces in the regeneration of post-industrial urban

[8] https://www.theguardian.com/cities/2017/oct/26/gentrification-richard-florida-interview-creative-class-new-urban-crisis.

landscapes. In practice, during the neoliberal era of the last three decades (Mayer et al. 2016; Rossi 2017), local authorities intensified their transfers of public assets to private corporations, welcomed investments in construction and speculation and renewed old city districts despite activists' contestation. These processes are responsible of the current housing crisis with its concomitant dynamics of gentrification, displacement and socio-spatial segregation.

Florida would probably advise his student that there is still room for running another private consultancy. But it is unlikely he would recommend looking at what urban movements are proposing and fighting for.

I do not mean that all social movements are right and contribute in a positive way to make better cities. But this has often been the case, especially when they engage in struggles to improve the lives of the most deprived and oppressed social groups. The provision of decent and affordable housing, demands for schools, health services and green areas, the self-management of community gardens and counter-cultural social centres (see Fig. 2.3 portraying a squatted social centre in Madrid, El Eko), the promotion of urban bicycling and the improvement of conditions for a free expression of migrants, ethnic minorities, women and LGBTIQ people in urban spaces, to name a few, are examples of the achievements that we should acknowledge.

Cities are more than entrepreneurs, as Florida suggests, but they are also more than the mere combination of demographic density and a socially diverse population, as the classic sociologists contended. Cities are essentially the result of forced residential moves; builders', cleaners', retailers' and many other inhabitants' work; struggles from the grassroots; and the specific interplays of economic, political and cultural structures. Class, gender and ethnic conflicts are at the core of the processes and power conflicts that make cities possible and real. Therefore, urban movements, quite often apart from the institutional realm of the state, are one of the fundamental expressions of those conflicts.

To conclude, I highlight a few bullet points based on previous research on urban activism and movements:

1. It has been observed that there is a strong presence of middle class and highly educated activists among urban movements. When working-class participants are also involved, leadership tends to be taken by middle-class members (Pickvance 1995, pp. 203, 210; Fainstein and Hirst 1995).

Fig. 2.3 Squatted Social Centre El Eko, Madrid (Spain), 2017 (*Source* Photograph by the Author)

2. Compared to other social movements, urban movements tend to share protest repertoires, but they can also perform very specific types of protest such as rent strikes, squatting and the proposal of alternative spatial plans (Pruijt 2007; Shephard and Smithsimon 2011, pp. 38–44). In particular, the squatting of buildings may entail many purposes and motivations related to

collective consumption, urban planning issues and grassroots re-appropriation of the urban voids left behind by real-estate speculators (Cattaneo and Martínez 2014; Martínez 2013).

3. On the one hand, urban movements tend to flourish in moments of crisis, especially in the phases of transition of national political regimes but also at the most acute peaks of economic turbulence such as the Global Financial recession of 2008 onwards, and so-called natural disasters such as earthquakes and floods whose devastation depends on very social, political and economic circumstances. These 'crises' are also felt in the periods of rapid economic growth, for example, in the aggressive renovations of old housing stock and the forced expulsion of long-term residents (in the last years, London, Moscow, Stockholm, Barcelona and Madrid, for instance, are examples of this[9]) (Baeten et al. 2017). On the other hand, the institutionalisation, co-optation and fragmentation of movements in their declining stages are not necessarily their final fate; they can mutate in form as well as last and continue at a different pace (Martínez et al. 2018b).
4. Urban movements focused on the redistribution of wealth and urban commons may intersect and overlap with other movements more focused on the right to the city, the self-management of spatial configurations and the contestation of either technocratic or limited participatory governance of cities. Both housing movements from Rome and Spain that I mentioned above are excellent representatives of these combinations.
5. Urban movements are attached to the local, but not exclusively (Hamel et al. 2000, pp. 1–12). There is a transnational nature in many locally bounded urban struggles (Mayer and Boudreau 2012, p. 284). Movements in particular have to deal with the politics of scale when authorities rescale resources, policies and decision-making processes, which, in turn, forces activists to 'multi-scalar strategies' (Nicholls et al. 2013, p. 9), as it is in the case of the Spanish housing movement.

[9] https://www.theguardian.com/society/2017/jul/21/the-real-cost-of-regeneration-social-housing-private-developers-pfi https://www.theguardian.com/cities/2017/oct/31/moscow-residents-vote-russia-demolition-rehousing https://www.elsaltodiario.com/vivienda/bajo-argumosa-11-la-playa.

In sum, far from understanding urban movements as irrational and spontaneous reactions to unbearable grievances, in the framework I have delineated these varieties of collective action are crucial and regular components of the production, governance and change of cities. Sometimes they succeed, and sometimes they fail, but the different interpretations about their outcomes should not overlook their specific role in urban politics. More equal, just, inclusive, democratic, balanced, sustainable, affordable, safe and liveable cities are not only the business of urban planners and municipal rulers. Indeed, these are quite prone to make decisions highly influenced by capitalist investors and the most privileged citizens. The basic pillars of liberal democracy are thus continuously threatened and undermined by private interests and authoritarian corporations where the main *leit motiv* is capital accumulation for a few. When urban movements with an emancipatory stance come to the fore—even by confronting elected authorities with their criticism and by breaking the conventional rules of institutional politics with direct actions—more participatory and genuine forms of democracy and economic redistribution are demanded and sometimes achieved. Collective action around the urban commons, rights and communities have shown extraordinary capacities to advance social cooperation among diverse groups and the self-management of crucial aspects of their lives. Therefore, I am convinced that only regular mobilisations from the grassroots may refresh the principles of a democratic city. Furthermore, urban movements, as the cases I mentioned, may counter the tendencies to far-right extremism, human rights violations, de-democratisation (Tilly and Tarrow 2007) and commodification (Merrifield 2014) that the rule of capital entails, especially during the current era of overwhelming flows of global financial speculation that substantially hit the spatial and social structures of urban life.

References

Andretta, M., Piazza, G., & Subirats, A. (2015). Urban Dynamics and Social Movements. In D. Della Porta & M. Diani (Eds.), *The Oxford Handbook of Social Movements* (pp. 200–218). Oxford: Oxford University.

Baeten, G., Westin, S., Pull, E., & Molina, I. (2017). Pressure and Violence: Housing Renovation and Displacement in Sweden. *Environment and Planning A, 49*(3), 631–651.

Carlsson, Ch. (Ed.). (2001). *Critical Mass. Bicycling's Defiant Celebration.* Oakland: AK Press.

Castells, M. (1983). *The City and the Grassroots. A Cross-Cultural Theory of Urban Social Movements.* Berkeley: University of California.

Cattaneo, C., & Martínez, M. (Eds.). (2014). *The Squatters' Movement in Europe: Commons and Autonomy as Alternatives to Capitalism.* London: Pluto Press.

Di Feliciantonio, C. (2017). Spaces of the Expelled as Spaces of the Urban Commons? Analysing the Re-emergence of Squatting Initiatives in Rome. *International Journal of Urban and Regional Research.* https://doi.org/10.1111/1468-2427.12513.

Fainstein, S., & Hirst, C. (1995). Urban Social Movements. In D. Judge, G. Stoker, & H. Wolman (Eds.), *Theories of Urban Politics* (pp. 181–204). London: Sage.

Hamel, P., Lustiger-Thaler, H., & Mayer, M. (2000). Introduction. Urban Social Movements—Local Thematics, Global Spaces. In P. Hamel, H. Lustiger-Thaler, & M. Mayer (Eds.), *Urban Movements in a Globalising World* (pp. 1–21). London: Routledge.

Harvey, D. (2016). *The Ways of the World.* New York: Oxford University.

Logan, J., & Molotch, H. (1987). *Urban Fortunes. The Political Economy of Place.* Berkley: University of California.

Marcuse, P. (2009). From Critical Urban Theory to the Right to the City. *City, 13*(2–3), 185–197.

Martí, M., & Bonet, J. (2008). Los movimientos urbanos: de la identidad a la glocalidad. *Scripta Nova XII*(270(121)).

Martínez, M. (2011). The Citizen Participation of Urban Movements. A Comparison Between Vigo and Porto. *International Journal of Urban and Regional Research, 35,* 147–171.

Martínez, M. (2013). The Squatters' Movement in Europe: A Durable Struggle for Social Autonomy in Urban Politics. *Antipode, 45*(4), 866–887.

Martínez, M. (2017). Squatters and Migrants in Madrid: Interactions, Contexts and Cycles. *Urban Studies, 54*(11), 2472–2489.

Martínez, M. (2018a). Bitter Wins or a Long-Distance Race? Social and Political Outcomes of the Spanish Housing Movement. *Housing Studies.* https://www.tandfonline.com/doi/full/10.1080/02673037.2018.1447094.

Martínez, M. (Ed.). (2018b). *The Urban Politics of Squatters' Movements.* New York: Palgrave Macmillan.

Mayer, M., & Boudreau, J. (2012). Social Movements in Urban Politics: Trends in Research and Practice. In P. John, K. Mossberger, & S. Clarke (Eds.), *The Oxford Handbook of Urban Politics* (pp. 273–291). London: Oxford University Press.

Mayer, M., Thörn, C., & Thörn, H. (Eds.). (2016). *Urban Uprisings: Challenging the Neoliberal City in Europe.* London: Palgrave Macmillan-Springer.

Merrifield, A. (2014). *The New Urban Question*. London: Pluto.

Mitchell, D. (2003). *The Right to the City. Social Justice and the Fight for Public Space*. New York: The Guilford Press.

Nicholls, W., Miller, B., & Beaumont, J. (2013). Introduction: Conceptualizing the Spatialities of Social Movements. In W. Nicholls, B. Miller, & J. Beaumont (Ed.), *Spaces of Contention. Spatialities and Social Movements* (pp. 1–23). Farnham: Ashgate.

Nofre, J., Giordano, E., Eldrdige, A., Martins, J. C., & Sequera, J. (2017). Nightlife, Tourism and Urban Change in the Quarter of Barceloneta (2001–2015). *Tourism Geographies*. https://doi.org/10.1080/14616688.2017.1375972.

Ortiz, S., & Gutiérrez, B. (2015). Planning from Below: Using Feminist Participatory Methods to Increase Women's Participation in Urban Planning. *Gender & Development, 23*(1), 113–126.

Piazza, G. (2011). "Locally Unwanted Land Use" Movements: The Role of Left-Wing Parties and Groups in Trans-Territorial Conflicts in Italy. *Modern Italy, 16*(3), 329–344.

Pickvance, C. (1995). Where Have Urban Movements Gone? In C. Hadjimichalis & D. Sadler (Eds.), *Europe at the Margins: New Mosaics of Inequality* (pp. 197–217). London: Wiley.

Pineda, C. (2013). Acapatzingo: construyendo comunidad urbana. *Contrapunto 3*, 49–60. https://www.academia.edu/11757688/Acapatzingo_construyendo_comunidad_urbana._El_Frente_Popular_Francisco_Villa-UNOPII.

Pruijt, H. (2007). Urban Movements. In G. Ritzer (Ed.), *Blackwell Encyclopedia of Sociology*. Malden: Blackwell.

Rossi, U. (2017). *Cities in Global Capitalism*. Cambridge: Polity.

Scott, J. (1990). *Domination and the Arts of Resistance. Hidden Transcripts*. Yale: Yale University.

Shephard, B., & Smithsimon, G. (2011). *The Beach Beneath the Streets. Contesting New York City's Public Spaces*. Albany: Excelsior-State University of New York.

Stavrides, S. (2017). *Urban Commoning in Autonomous Neighborhoods of Mexico City*. Paper Presented in the RC21 Conference, Leeds.

Sweet, E., & Ortiz, S. (2010). Planning Responds to Gender Violence: Evidence from Spain, Mexico and the United States. *Urban Studies, 47*(10), 2129–2147.

Tilly, C., & Tarrow, S. (2007). *Contentious Politics*. New York: Oxford University.

UN-Habitat. (2016). *Habitat III Thematic Meeting on Informal Settlements*. Nairobi: UN-Habitat.

Villasante, T., et al. (1989). *Retrato de chabolista con piso. Análisis de redes sociales en la remodelación de barrios de Madrid*. Madrid: IVIMA-SGV-Alfoz. http://oa.upm.es/14695/2/Retrato_de_chabolista_con_piso_2.pdf.

PART II

Changing Forms of Urban Activism

CHAPTER 3

Housing Activism Against the Production of Ignorance: Some Lessons from the UK

Tom Slater

Introduction: Unsettling Supply and Demand

Housing—having a roof over one's head—is absolutely central to human dignity, community, family, class solidarity and life chances (Madden and Marcuse 2016). But intersecting with draconian welfare reforms, housing policies in the UK (particularly but not exclusively in England) are wreaking havoc upon people living at the bottom of the

This chapter draws materials from my research which have also been published in the following articles: Slater, T. (2018). The Invention of the 'Sink Estate': Consequential Categorisation and the UK Housing Crisis. In I. Tyler & T. Slater (Eds.), *The Sociology of Stigma*. London: Sage; Slater, T. (2016). Revanchism, Stigma and the Production of Ignorance: Housing Struggles in Austerity Britain. *Research in Political Economy, 31*, 23–48; Slater, T. (2016). The Housing Crisis in Neoliberal Britain: Free Market Think Tanks and the Production of Ignorance. In S. Springer, K. Birch, & J. MacLeavy (Eds.), *The Routledge Handbook of Neoliberalism* (pp. 370–382). London: Routledge.

T. Slater (✉)
University of Edinburgh, Edinburgh, UK
e-mail: tom.slater@ed.ac.uk

N. M. Yip et al. (eds.),
Contested Cities and Urban Activism, The Contemporary City,
https://doi.org/10.1007/978-981-13-1730-9_3

class structure. A few snapshots of the situation across the UK suffice to assess the financial ruin and displacement of the poor created by four decades of housing policies tightly tethered to profit generation for owners of land and property and correspondingly unmoored from providing shelter for people most in need. As Lansley and Mack (2015) detail in *Breadline Britain*, of the four million people in the private-rented sector (PRS) who live in poverty, a full two million of those are employed full time. One-third of all private-rented sector tenants in the UK are living in structurally inadequate housing, with poor insulation issues having major energy and health implications. More than two million households (and counting) are on the waiting list for social housing. A staggering 1.8 million households are spending over half their incomes on housing costs: the very poorest people have approximately £60 per week left for everything after housing costs are met. Local authorities have spent £3.5 billion on temporary housing in the last five years (Buchanan and Woodcock 2016). Homelessness has become a fixture of cities and is still on the rise (there has been a substantial increase in rough sleeping since 2010)—even though there are over 750,000 empty homes across the UK. Security of tenure is a huge issue, amplified by the massive rise in 'assured shorthold tenancies' because of the explosion in 'buy-to-let' mortgages, a get-rich-quick scheme for landlords that until 2015 offered generous tax breaks, and still allows landlords to evict tenants without any reason. If food prices had risen to the same rate as house prices since 1971, a fresh chicken would cost over £50 (Carylon 2013). Under the banner of 'regeneration', social housing in English cities, particularly London, is being demolished at an unprecedented rate without replacement (Watt and Minton 2016). Perhaps most arresting of all is that one-third of Conservative MPs have vested interests in maintaining the status quo, for they are private sector landlords. Profit has been the guiding principle behind government housing policies for four decades. Spectacular fortunes have been made, but the cocktail of deregulation, privatisation and attacks on the welfare state has also made a spectacular mess.

Against this backdrop—and particularly against the 2015 General Election backdrop of the Labour Party taking high housing costs more seriously than in its recent history and proposing an upper limit on rent increases within tenancies in the PRS—the *Institute of Economic Affairs* (IEA), a free market think tank and a pivotal institution in the birth of neoliberal ideology in the UK and beyond, published a report entitled *The Flaws in Rent Ceilings*[1] (Bourne 2014). A declamatory crusade against all forms of rent regulation anywhere, the report began by stating that there is a 'rare consensus' among economists that rent control 'leads to a fall in the quantity of rental property available and a reduction in the quality of the existing stock' (Bourne 2014, p. 10). A quick inspection of a footnoted URL reveals that the source of this 'consensus' is in fact a small survey of 40 neoclassical economists in the USA. Nonetheless, the report continued to argue that 'under rent control there is less incentive for families to reduce their accommodation demands, therefore exacerbating the shortage of properties for others' (Bourne 2014, p. 16). The tenor of the document reaches a crescendo a few pages later in the spectacular assertion that 'the truth would appear to be that tenants are unwilling to pay for increased security' (Bourne 2014, p. 25), leading to the conclusion that any 'extra security' for tenants 'comes at the expense of reduced economic efficiency' (Bourne 2014, p. 35). Instead of rent regulation, the report calls for another round of deregulation in the form of 'planning liberalisation', which is described as a 'welfare enhancing policy' (Bourne 2014, p. 36) that would lead to the construction of new housing on land currently shielded from development by government red tape. For the IEA, the housing crisis is a basic economic conundrum—too much demand and not enough supply—and its solution is thus to increase supply by stopping all government interference in the competitive housing market, which (true to neoclassical beliefs) must be allowed to operate free of cumbersome restrictions to provide incentives for producers and consumers to optimise their behaviour and push the market towards equilibrium (so that there are no shortages of housing), whilst yielding the maximum amount of utility for the maximum number of people.

The IEA immediately went about the task of circulating sound bites from the report as widely as possible. Its 'solution' certainly caught the attention of newspapers and commentators supporting a conservative agenda, one illustration being the *The Daily Telegraph* printing a feature under the headline, 'Think-tank criticises "pointless" Labour

[1]This is not the first time the IEA has waded into the debate on rent control. Founded in 1957 by Anthony Fisher, an ex-RAF pilot, wealthy chicken farmer and personal friend of Friedrich von Hayek, in 1972 the IEA published a savage tirade against state intervention housing markets entitled 'Verdict on Rent Control', co-authored by none other than Hayek himself with another grandfather of neoliberalism, Milton Friedman.

rent cap scheme' (6 September 2014). It also caught the attention of the editors of *Channel 4 News*, a widely respected national television news programme, who invited the author of the report, Ryan Bourne (then the IEA's Head of Public Policy), to discuss the issue of rent control alongside Jasmine Stone,[2] an activist from the London FOCUS E15 movement (which campaigns against the demolition of social housing). Even after Stone described her and her neighbours' experiences of struggling to make rent due to various profiteering schemes, Bourne maintained that the 'fundamental reason' for very high rents was not 'greedy landlords'. According to Bourne, the 'real' reason is that, 'Over years and years we haven't built new homes and have restricted supply artificially[3] through greenbelts and other planning restrictions'. When Stone answered positively to the interviewer's question of whether she would like to see rent controls introduced in London, Bourne immediately retorted that she was wrong because 'economists agree' that such 'crude' controls are 'absolutely disastrous'. It did not appear to matter at all to Bourne that Stone was speaking from the experience of poverty, housing precarity and repeated evictions.

This chapter traces the historical development of this unequal context and explores in particular the organising work of the Living Rent campaign in Scotland, which has achieved significant policy changes in Scotland in a very short space of time, largely due to a canny mix of direct action, dialogue with key politicians and alternative knowledge production against the production and circulation of ignorance in respect of housing affordability and rent control. I argue that, notwithstanding remarkable progress, tenant movements such as Living Rent still have to confront the ignorance that is intentionally produced by the powerful institutions behind what we might call vested interest urbanism. Correspondingly, I situate my analysis within the register of *agnotology*, a body of work that has emerged to expose and critique the intentional production of ignorance.

A crucial geographical caveat is necessary before I proceed: there are significant spatial and national differences in the political and social landscapes of Scotland and England especially. Kim McKee, a prominent

[2] Available to view here: http://t.co/ZLzoiHp2Su.

[3] The use of the word 'artificially' is revealing: neoclassical economists are especially prone to viewing a competitive market as a natural evolution that is best left alone if equilibrium and growth is to be achieved.

housing scholar in Scotland, recently argued, 'Given the devolved nature of public policy making in the UK it is no longer possible to talk of a British experience, if indeed it ever was' (McKee 2016, p. 2). She is absolutely correct on housing policy (and also education, health care, etc.), but then there is welfare reform, which is UK government legislation that is not devolved and which has huge implications for housing and for cities. Therefore, I retain 'Britain' as a frame of reference, not only because of welfare reform but also because of the capitalist accumulation motives behind different policy approaches and because the circulation of interest-bearing capital in urban land markets—so crucial in understanding the high cost of housing—tends not to recognise national borders.

A Brief Note on Agnotology

> It is certain, in any case, that ignorance, allied with power, is the most ferocious enemy justice can have.
>
> James Baldwin (1972)

In his swashbuckling critique of the economics profession in the build-up to and aftermath of the 2008 financial crisis, Mirowski (2013) argues that one of the major ambitions of politicians, economists, journalists and pundits enamoured with (or seduced by) neoliberalism is to plant doubt and ignorance among the populace:

> This is not done out of sheer cussedness; it is a political tactic, a means to a larger end. ...Think of the documented existence of climate-change denial, and then simply shift it over into economics. (Mirowski 2013, p. 83)

Mirowski makes a compelling argument to shift questions away from 'what people know' about the society in which they live towards questions about what people do *not* know and why not. These questions are just as important, usually far more scandalous and remarkably undertheorised. They require a rejection of appeals to 'epistemology' and, instead, an analytic focus on intentional ignorance production or *agnotology*. This term was coined by historian of science Robert Proctor, to designate 'the study of ignorance making, the lost and forgotten' where the 'focus is on knowledge that could have been but wasn't, or should be but isn't' (Proctor and Schiebinger 2008, p. vii). It was whilst investigating the tobacco industry's efforts to manufacture doubt about the health hazards of smoking that Proctor began to see the scientific and political urgency in researching how ignorance is made, maintained and manipulated by powerful institutions to suit their own ends, where the guiding research question becomes, 'Why don't we know what we don't know?' As he discovered, the industry went to great lengths to give the impression that the cancer risks of cigarette smoking were still an open question even when the scientific evidence was overwhelming. Numerous tactics were deployed by the tobacco industry to divert attention from the causal link between smoking and cancer, such as the production of duplicitous press releases, the publication of 'nobody knows the answers' white papers and the generous funding of decoy or red-herring research that 'would seem to be addressing tobacco and health, while really doing nothing of the sort' (ibid., p. 14). The tobacco industry actually produced research about everything except tobacco hazards to exploit public uncertainty (researchers commissioned by the tobacco industry knew from the beginning what they were supposed to find and not find), and the very fact of research being funded allowed the industry to say it was studying the problem. In sum, there are powerful institutions that want people not to know and not to think about certain conditions and their causes, and agnotology is an approach that traces how and why this happens.

Many scholars (and think tank writers) might claim that there is no such thing as the intentional production of ignorance; all that exists are people with different world views, interests and opinions, and people simply argue and defend their beliefs with passion. Yet as I will demonstrate with reference to the debate on rent control, this claim would be very wide of the mark. Even when there is a vast body of evidence that is wildly at odds with what is being stated, and when the social realities of poverty and inequality expose the failures of deregulation at the top and punitive intervention at the bottom of the class structure, the 'free marketeers' become noisier and even more zealous in their relentless mission to inject doubt into the conversation and ultimately make their audiences believe that government interference in the workings of the 'free' market is damaging society. Therefore, tracking the ignorance production methods of 'the outer think-tank shells of the neoliberal Russian doll', to use Mirowski's (2013, p. 229) memorable phrasing, is a project of considerable analytic importance.

The Housing Crisis and the Rise of the Private-Rented Sector

In 2010, a Coalition government came to power in Britain following an election which did not yield a clear majority for any party. That Coalition was a skewed alliance between the dominant Conservative Party (which campaigned using the language of compassion and social progress to shield the electorate from its rabidly right wing, ruling class and corporate ethos) and the subordinate Liberal Democrats (a small set of centre-right political lightweights without a coherent message or set of policies, nearly obliterated at the polls in 2015). David Cameron, and his Chancellor of the Exchequer, George Osborne (both members of the British aristocracy with substantial family fortunes, who have surrounded themselves with many more such people), arrived in office during a global financial crisis and were confronted by a substantial budget deficit which they argued was a consequence of irresponsible public spending by the previous Labour government, rather than what it actually was: the result of massive government bailouts of financial institutions responsible for the crisis. For Cameron and Osborne, two archdeacons of low taxation and low public spending, there was only one way to deal with this budget deficit: a vicious austerity package, which, conveniently, was also an opportunity to finish off the welfare state that 'Thatcher's children' of the Conservative Party so despise and replace it with their dream of a thoroughly privatised and individualised society which would protect the sanctity of private property rights and a free market.[4] To continue the relentless pace of expanding global accumulation, British ruling elites have set out to monitor and monetise more and more of those human needs that were not commodified in previous rounds of financialisation. Pensions, health care, education and especially housing have been more aggressively appropriated, colonised and financialised (Meek 2014). For Conservatives, the redistributive path—increasing taxation of corporations, land and property (London in particular is known as a tax haven for foreign investors in land and property)—is not a matter for public

[4] In case there is any doubt about intentionality, it is worth remembering that in April 2011 the Conservative MP Greg Barker (then Minister of State for Energy and Climate Change) told an audience in the USA: 'We are making cuts that Margaret Thatcher, back in the 1980s, could only have dreamed of'. http://politicalscrapbook.net/2011/04/greg-barker-thatcher-video/.

discussion, and an entire cadre of cultural-technical experts (chief among them economists, lawyers, think tank researchers and communications professionals) is in place to make sure the conversation does not head in that direction. This ensures that it is largely unknown that an estimated £120 billion a year is lost in the UK due to corporate tax avoidance, evasion and collection errors.[5] The Tories were re-elected with a majority in 2015, but the Brexit fiasco of 2016 saw Cameron's resignation and his replacement with Theresa May, who has continued the Tories' austerity agenda against growing opposition and a resurgent Labour Party under Jeremy Corbyn.

Whilst welfare 'reform' (demolition would be a better characterisation) in the UK encompasses far more than just housing, it is the cuts to housing benefit in particular that have become the flashpoint for a compelling and disturbing class struggle. Housing benefit in Britain was introduced in 1988, and its objective was to subsidise the cost of rental accommodation for tenants in the social-rented and private-rented housing sectors (Hamnett 2010). It is a means-tested benefit, which is administered by local authorities and paid to eligible tenants (to be eligible, a household's income has to be less than £16,000 a year). Entitlement is calculated by comparing current household needs and resources with household rent payments. Nearly five million people (about 10% of all adults in Britain) claim an average weekly rental subsidy of £84—in London, the most expensive city, the average climbs to £130 (Hamnett 2010). Recognising the uneven geography of housing costs, the last Labour government in 2008 introduced something called the Local Housing Allowance, for new claimants living in the private-rented sector: a flat-rate allowance for properties of different size in local rental markets. This scheme led to very high levels of payment in a very small number of cases, up to £2000 per week for a four-bedroom house in central London—entirely because of the scandalously high cost of housing there (which has nothing to do with the housing benefit claimant). Nonetheless, since 2010 the Conservatives have argued that it is 'unfair' that a 'hardworking household' is, via their tax payments,

[5] Just the money through tax avoidance alone could pay for 25,000 nurses on a £24,000 a year salary for 20 years; could put 129,000 children through school from ages 5–18; and would allow the government to give every single pensioner in the UK an extra £65 a year: http://leftfootforward.org/2013/09/another-crackdown-on-benefit-fraud-yet-it-accounts-for-just-0-7-per-cent-of-welfare-budget/.

enabling a household where nobody works ('a feckless, work-shy household', according to Iain Duncan-Smith, infamously modified by George Osborne to 'shirkers') to live on the same street or same neighbourhood and sometimes to live in a larger house in a more desirable district. Therefore, the government has capped housing benefit at £400 per week for a four-bedroom home, £340 per week for a three-bed, £290 per week for a two-bed and £250 per week for a one-bed.

Why is housing benefit being targeted like this? Notwithstanding right-wing critiques of the implications of government 'handouts', the main reason is that the annual housing benefit bill is enormous, up from £six billion annually in the mid-1990s to nearly £40 billion today: an easy target for a government wedded to austerity. But there are crucial historical reasons why this bill is so high that are conveniently ignored by those attacking housing benefit, which have everything to do with the explosion of the PRS in the UK. This is a direct outcome of Margaret Thatcher's game-changer: the infamous Right to Buy scheme,[6] introduced in 1980 to encourage direct sales of council housing at large discounts to tenants (to promote a 'property-owning democracy' at the expense of the public housing stock). Over the past 35 years, nearly three million publicly owned homes have been sold off under the scheme. Many working-class people acquired an asset beyond their dreams, at a steal. Yet Right to Buy actually failed on its own privatising terms, as over 40% of all those who exercised their Right to Buy then sold on to private landlords (Collinson 2017), who rented them to tenants at double or triple the levels of council rent, *which required tenants to apply for housing benefit off the state*. So Thatcher's flagship policy actually ended up costing the state far, far more in housing benefit than it ever did in the maintenance and management of council homes. None of this appears to matter to the Conservative government: Right to Buy has been extended to 1.3 million tenants in housing association properties, at discounts of more than £100,000 per property in London and £70,000 elsewhere.

[6]RTB has been the largest but by no means the only economic deregulation and privatisation scheme vis-à-vis housing in Britain. Many remaining council housing stocks have been transferred to housing associations ('stock transfer') as part of 'the state's desire to end municipal landlordism altogether, accounting for a further 1.5 million public dwellings' (Hodkinson 2012, p. 510). Year upon year, there is a dwindling public subsidy for housing associations, which have been forced to rely on commercial borrowing to balance the books, putting upward pressure on rents (Hodkinson 2012).

Without question, the main long-term effect of Right to Buy was to rob future generations of affordable housing, diverting them into the landlord bonanza that is the private-rented sector. Since 1979, there has also been a substantial reduction in the amount of new social housing construction, which has led to the steady shrinkage of the sector. Senior Tories have even spoken of their desire to rebrand social housing as 'taxpayer subsidised housing'.

It doesn't take a complex modelling exercise to conclude that the high cost of housing together with the assault on the welfare state means that many thousands of people of low incomes are no longer able to pay rent in cities across the UK, particularly those living in cities where the cost of housing is so exorbitant. This is leading to displacement on an epic scale, especially from London. In April 2015, it was revealed by reporters at *The Independent* newspaper that local authorities in London moved almost 50,000 families who had fallen into rent arrears out of the city between July 2011 and July 2014—sometimes at a rate of nearly 500 families a week. In many cases, families were moved to cities faraway from social networks in London: to Manchester, Bradford, Hastings, Dover and Plymouth, among others. When this is considered alongside the further cuts that were introduced in April 2016, where the *total* benefits a family can receive (not just in housing) were reduced from £26,000 to £20,000 (£385 a week) and to £23,000 (£442) in London, large parts of the south-east and south-west become unaffordable. Figure 3.1 offers a cartographic depiction of the implications for the case of England, showing the areas where a family of two adults and two children living in a two-bedroom home will be hit by the benefit cap.

The displacement of working-class people from central city districts in Britain, particularly but not exclusively from districts in London and south-east England, should not be viewed as a consequence of the way the housing market works via the 'laws' of supply and demand, nor as a 'natural evolution' (Ball 2014), but as calculated expulsion being engineered by a revanchist national state in alignment with the interests of private capital. A principal motive is the capture of rent from highly lucrative urban land markets. The crisis of affordable housing in the expanding PRS points to the absolute necessity of rent regulation to protect increasing numbers of people being hammered by a housing market that offers frequently staggering returns only to those with the financial means to compete (not to mention hammered by a hostile, precarious entry-level labour market into which the most vulnerable are being

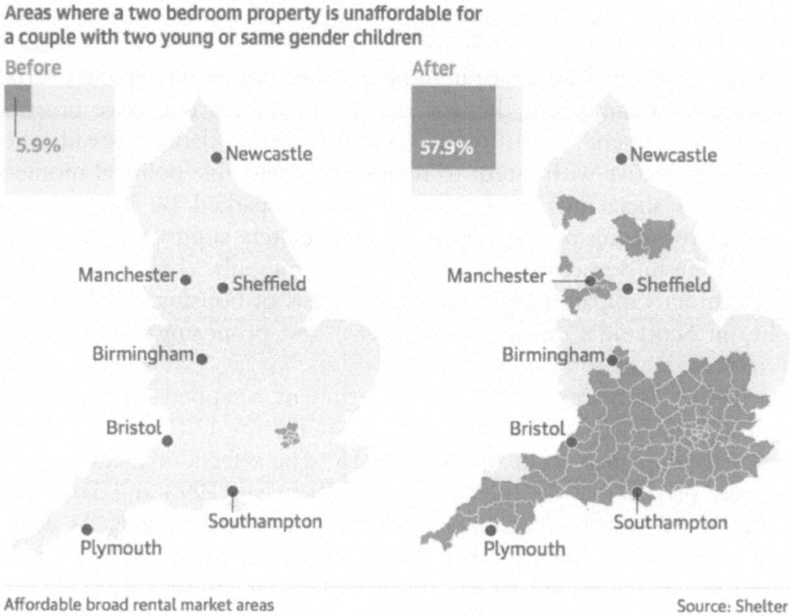

Fig. 3.1 Housing affordability in England before and after the assault on housing benefit (*Source* Pennington (2015) in collaboration with *Guardian* newspaper. Reproduced with permission from Shelter)

funnelled by workfarist reforms). Housing activists in Scottish cities recognise this necessity, and their recent organising around the question of rent control warrants close attention.

A Living Rent?

The 'Living Rent' campaign[7] was formed in the aftermath of the Scottish Independence Referendum in 2014. Its leadership is comprised entirely of people in their twenties—the age cohort hardest hit by high housing costs in rental accommodation. Between 2003 and 2014, the proportion of 16- to 24-year-olds whose rent privately increased from 30 to 59%, and the proportion of 25- to 34-year-olds renting privately almost

[7] http://www.livingrent.org/.

tripled, from 13 to 35% (Maloney 2015). All of the key figures in the campaign have experienced issues with rapacious/unethical landlords and have been involved in a range of housing justice movements across Scotland. As a campaign, the activists do not take a collective position on the ever-present constitutional question in Scotland; instead, they formed a collective with intent to retain the remarkable political momentum around social justice issues in Scotland sparked by Referendum. Although the name of the campaign immediately suggests links to the Living Wage movement, these links are in name only: they are clear that this a distinct campaign highlighting the crisis of housing costs housing quality in Scotland's private-rented sector and proposing solutions and alternatives.

In 1988, Margaret Thatcher's government scrapped rent controls, which had been in place in various forms across the UK since the watershed of the Glasgow Rent Strike of 1915. The effects of this deregulation were predictable. In the two decades between 1993 and 2013, the private-rented sector in Scotland more than doubled in size. As well a direct consequence of 'buy-to-let' investment, this expansion must be understood as correlated to the decline of social housing from 34% of the housing stock to just 13% (particularly via demolitions and 'stock transfer') and to rapidly increasing house prices in the owner-occupied sector that present a major barrier to first-time buyers (in the 12 months leading up to March 2015, house prices increased faster in Scotland than anywhere else in the UK, and more than 3% faster than London) (Maloney 2015). Additionally, the gender pay gap in Scotland sees women earn as much as 17.8% less than men, which means that high rents have a disproportionate impact on women, especially single parents. The Living Rent activists are not only concerned with high costs of private renting, but very poor standards. More than half of all dwellings in the sector fail the Scottish Housing Quality Standard, with structural defects, rising damp and lack of insulation being critical problems affecting the respiratory health of tenants (leading to indirect health care costs). Another issue is tenant security. Scottish tenants are among the least secure in Europe: 'exceptional in offering tenants very short periods of security from eviction and no protection against rent rises for the duration of the tenancy' (Maloney 2015, p. 9). Eviction rates have never been higher (as is the case in England and Wales) and will only increase given the impact of welfare reform. Without question, the situation is one of crisis, and Living Rent has embarked on a campaign to address it.

The campaign began by reaching out to existing organisations, particularly workers and student unions, and to the general public, notably through a large number of weekend street stalls and online activism. Living Rent also felt that reaching out to tenants directly was a form of disruptive action, as they are often not a constituency that is consulted and/or has the means to put its voice across, and thus, it is often outside organisations (usually with tenant issues in background) who speak for them. It also organised marches and public protests to deliver completed petitions, and a major campaigning intention was to push the ruling Scottish National Party (SNP) to be more progressive about housing issues given the overwhelming support for progressive policies that had inspired the YES vote in the independence Referendum. They accomplished this notably through passing motions within the more progressive and young branches of the SNP and providing a well-researched policy brief to key SNP politicians.

Within six months of beginning their campaigning, Living Rent had affiliations with organisations representing more than a million Scots, from trade unions and student associations' to women's organisations, faith and youth groups and more. This is a reflection of Living Rent was able to articulate the intersection of housing precarity with other social problems—how housing affects so many different groups in particularly acute ways. These organisations backed the campaign because 'rent control', quite simply, resonated with them all as a common-sense solution to an increasingly pressing problem: the high cost of housing consigning people to financial ruin. Living Rent had a strong commitment to accomplish a very simply mission: to bring back the words 'rent control' to the political debate where it had been erased for so long. The leaders of the campaign also worked with critical urban scholars in Scotland to research different models of rent controls throughout the world to understand which models 'worked' and which didn't, with a view to outlining what kind of model Living Rent wanted. The campaign was particularly successful in taking academic work and translating it into bite-size sound bites that people could connect with, and that politicians were asked to back publically. Social media was also crucial, bringing radical ideas into other forums such as facebook, twitter and online petitions.

In their conclusion to *Contesting Neoliberalism*, a book with an admirably broad sweep of case studies across the globe, Leitner et al. (2007) identified four 'trajectories' followed by various movements and struggles fighting against the mobilisation of state power in the extension of market rule:

1. *engagement*, whereby non-neoliberal interests opt for (or see no alternative to) cooperation with neoliberal corporate and institutional power,
2. *direct opposition* to neoliberal corporate and institutional power (via various tactics of organised collective action),
3. *alternative knowledge production*, which involves decentring neoliberal truth claims by disseminating alternative interpretations, facts and arguments, and
4. *disengagement*, or developing spaces within which alternative practices can be pursued in their own right and on their own terms.

What is striking about Living Rent is that it has combined the first three trajectories to remarkable effect in a very short space of time. In respect of *engagement*, campaign leaders saw an opportunity to intervene when the Scottish Government announced a consultation on housing affordability problems across all sectors. Given the dominance of Scottish politics by the SNP, key SNP politicians were targeted for constructive dialogue and educated about housing problems and about the cost savings to the state from rent regulation in particular. In respect of *direct opposition*, Living Rent holds demonstrations and rallies outside the offices of notorious letting agents in Scottish cities and continues to organise marches to coincide with key occasions and debates in the Scottish political calendar and protests against various factions of the property industry in Scotland. In respect of *alternative knowledge production*, this is where Living Rent has arguably been most impressive: a central goal has been to unmask and challenge myths about rent control and use its website as a clearing house of information on this issue, complete with useful facts and statistics, guest blogs and policy briefs. Blending these trajectories, Living Rent called for the following:

- For initial rents to be set against a 'points' system to reflect the value of the property (using rent regulation in the Netherlands as a model).
- For rent increases to be capped at a rent affordability index to ensure increases do not push tenants into hardship.
- For a move towards indefinite tenancies as default, away from short-term contracts.
- For all tenants to be entitled to a hardship defence in relation to evictions.

- For the creation of a Scottish Living Rent Commission, to oversee these recommendations and to serve as a centre of expertise for the Scottish Private Rental Sector.

The resulting achievements are truly impressive, given over three decades of the rampant neoliberalisation of housing in Scotland. Not only they have built alliances with (notoriously sectarian) labour unions to widen their audience, but, in response, the Scottish Government announced the introduction of rent controls in certain 'rent pressure areas' in Scottish cities, starting in April 2016, and also announced the end of the right of landlords to reclaim their properties from tenants without any reason. Although these announcements do not go anything like deep enough for Living Rent, the question of rent control was indeed *not even on the political agenda* in Scotland before the activists came together. Their plans now are to keep up the campaign for much stricter rent controls (the formula for designating a 'rent pressure area' does not link the price to the quality of the home or its energy efficiency, but only sets the rent of a home to the price laid out by the market; the controls introduced just seek to manage the increases rather than freeze or lower rents); to make the campaign more grassroots (including more direct action); and to build a tenants' union to shift the power balance away from landlords and towards the question of the human need of shelter. In what follows I outline some arguments against the mythology that rent controls do not work, which have been developed with my colleague at Edinburgh, Hamish Kallin, working with Living Rent activists.

Challenging the Production of Ignorance on Rent Controls

There are three myths that are replicated again and again in the arguments against rent controls: that they threaten the quality, supply, and efficiency of the housing sector. The quality argument goes as follows: rent controls would negatively affect the standard of homes on offer, so if a landlord cannot raise rents as much as they want to, they are likely to skimp on maintenance, or—even worse—have insufficient funds to carry out necessary maintenance, even if they wanted to. The most obvious flaw with such an argument is that housing quality within the private-rented sector is already atrocious. Indeed, it is the worst of all tenures, with the most sophisticated surveys of poverty, showing that one in three tenants in the PRS lives in structurally inadequate housing (Lansley and Mack 2015). This is confirmed by the government's own reports: almost one-third of privately rented accommodation fails to meet the government's standards for decent homes (UK Parliament 2016). The housing charity Shelter (2014, p. 7) reports that over six in ten renters (61%) have experienced at least one of the following problems in their (privately rented) home over the last 12 months: damp, mould, leaking roofs or windows, electrical hazards, animal infestations and gas leaks. Ten per cent of renters said their health had been affected because of their landlord not dealing with repairs and poor conditions in their property in the last year, and nine per cent of private-renting parents said their children's health had been affected. Moreover, in the decades before state intervention in housing (when the vast majority of the UK population were privately renting), standards were far worse. The historical record of 'laissez-faire' liberalism on housing standards was simply terrible, with slum conditions and overcrowding commonplace in British cities, where chronic poverty would siphon wealth upwards through rent (Engels 1845; Rodger 1989). Those arguing that rent controls would worsen housing quality cannot have it both ways: whenever there has been little or no regulation, rental housing quality has been appalling.

Overcrowding in Scottish cities in the early twentieth century (before the 'red tape' of state regulation interfered) was chronic, housebuilding was inefficient, and evictions were common (Rodger 1989). Glasgow in 1900 was as close to the conditions of a 'perfect free market' in housing as the likes of Niemietz could possibly desire—no public housing, no regulated standards of accommodation, a lack of monopoly in the hands of any single owner and virtually no protection of tenants' rights. But rents were high and conditions were appalling (McCrone and Elliot 1989), with slum landlords cramming tenants into stairwells, courtyards, alleys—denying them access to light, water, or dignity (Gauldie 1976). The 'luxuries' of fire safety, running water, central heating, indoor toilets, watertight roofing and so on were won through political struggle over the decades that followed and only normalised through legislation. And, as the fire at Grenfell Tower in the summer of 2016 so horrifically demonstrated, the 'profit motive' cuts corners wherever it can. Only regulation—effectively enforced—can uphold decent housing standards. It is frankly absurd to say that introducing even modest rent regulation would make that quality problem worse. This is abundantly clear in the case of the Netherlands, where the amount a landlord is allowed to increase rent on an annual basis is conditional upon the standard of the property they are leasing. The collective result is a rental housing stock in far better shape than in countries that have no rent control (Olsen 1988; Anas 1997; Kutty 1996).

The second myth concerns the question of supply. The hegemonic view is that the housing crisis is a simple imbalance between supply and demand. Were state restrictions lifted, we are promised, supply would increase, and prices would fall. But the problem with such a formula is that it overlooks the myriad reasons for this crisis which are not caused by legislative barriers to profit, but rampant speculation itself, whilst remaining blind to rising levels of inequality. Moreover, as Minton (2017) reminds us, the 'top end' of the housing crisis is characterised by hyper-investment in the most expensive parts of the city. 'Demand' in this context bears no rational relation to 'supply' because the housing is not being bought to be lived in, but purely as a financial asset. Since mid-2007, the supply and demand narrative has been shown to be 'wholly unsatisfactory' (Bone and O'Reilly 2010, p. 234) because it cannot explain the role of financialisation and the rise in prices far, far beyond levels of affordability. Research by the estate agent Savills (2017) also (unwittingly) demonstrates that 'supply' follows the money, not the need: in London, there is a surplus of housing in the upper end of the market—'ready for increased international demand' (Savills 2017, p. 5) to use their positive spin—but as you work your way down the affordability ladder the surplus-to-demand ratio diminishes to the point of crisis. The idea that building more in the upper end will help those with less money is a fantasy. The opposite is now demonstrably true. It is the price inflation at the upper end that 'trickles down' the housing ladder, not the surplus (Minton 2017). Furthermore, housing wealth does not trickle down; it flows upwards, into the hands of rentiers, lenders, investors, speculators, developers and housebuilding corporations. Put bluntly, the logic of supply and demand is useless to explain the housing crisis, because people demand homes to live in, but the driving force of financialisation ensures that any link between what a home is used for and what it is sold for has been utterly broken (Rolnik 2013).

Were rents to be capped, we are told, then fewer people would bother to become landlords, existing landlords would withdraw their properties from the market, and fewer developers would bother to build. The result would be a restriction in the supply of new housing for rent, which would lead to the housing crisis getting even worse. This is neoclassical logic writ large and a profoundly troubling argument for two reasons. First, it is an unconscious admission that the PRS is only viable so long as it can exploit people far beyond their means—a formula based on parasitic greed. Second, it implies that any curtailing of the profits to be made from a sector will simply stop people investing in it. This is akin to believing that the minimum wage means companies stop employing people; that VAT means nobody buys anything any more; that Fuel Duty means nobody drives nowadays; it is, in other words, a fantastical formula, for it is based on the notion that people will only seek to make money in conditions of totally unhampered profitability. Such a hypothesis is, once again, an ahistorical utopia (in the sense that it has never existed). The precipitous decline of the PRS in Britain over much of the twentieth century—from near 90% at its start down to 14% by the 1970s (Stafford 1976, p. 3)—was clearly influenced by blanket rent controls (not what anyone is arguing for today), but it would be simplistic to say that rent controls alone led to this decline. Moreover, this housing did not disappear, it simply changed tenure (either because more people were living in state-owned housing or, following privatisation, as owner-occupiers). To then suggest that any decline of the PRS in and of itself is a crisis seems to be misplaced. Similarly, it is inaccurate to suggest that the massive expansion of the PRS in recent decades has been caused solely by the abolition of rent regulation. It is well documented that decades of drastically reducing the affordable housing stock alongside the stagnation of wages and the rise in house prices have left people with few alternative options (Meek 2014). In other words, the growth of the PRS looks more like a symptom of the current housing crisis than it does its solution. If we push this argument, it stands to reason that we start to question the existence of the PRS itself. If the most strident critique of rent controls is that it will lead to a reduction in the sector, it is sorely tempting to respond by saying, 'Sounds good!' As an innately exploitative relation, private renting is a parasitic form of accumulation which funnels wealth up. It fuels rising inequality and lends it a cruel generational permanence. In this light, a downsizing of the PRS can hardly be considered a humanitarian tragedy if it makes decommodified forms of housing provision an urgent necessity. The 'supply' argument only makes sense if it denies all other forms of housebuilding and tenures and ignores the possibility of collective ownership.

Finally, we come to the holy grail of 'efficiency', a word that can be interpreted in so many ways it begins to feel thin. Through the frame of neoclassical economics, something is inefficient if it 'artificially' interferes with the 'natural' operation of the price mechanism of the market. Rent control, therefore, is viewed as a form of price fixing, which will have deleterious consequences in terms of encouraging the problem of 'sitting tenants' who will (a) block outsiders to the rental property market from gaining a foothold in it and (b) affect the functioning of a 'dynamic' labour market as they will refuse to move house to take any offer of employment elsewhere (as they would have to give up their low-cost rental housing if they did). The problem with this efficiency argument is not only the quite breathtaking assumption that low-income consumers have the freedom to 'rationally choose' where they want to live, without any kind of structural constraints in their lives, but the way it is skewed towards the interests of landlords, such as the IEA's 'landlords in a secure tenancy framework would face the prospect of "problem tenants" enjoying greater security of tenure, making the management of risk through turnover more difficult' (Bourne 2014, p. 25). Language matters: the 'management of risk through turnover' is a very polite way of describing evictions, and a 'problem tenant' is someone who cannot afford to pay the rent. If we are interested in housing as a question of social justice, then 'economic efficiency' arguments are to be treated with the utmost caution. Real 'efficiency' is surely achieved when rental housing costs are way, way lower than 50% of household incomes, when households have more money to spend on other necessities (and luxuries) and when the state does not have to haemorrhage £40 billion a year on subsidies to private landlords through housing benefit. In choosing between fears over price fixing and concerns for human well-being, the basis of a humane and sensible housing policy must side with the latter.

Conclusion

Rent regulation is an urgently required short-term measure to protect against evictions and homelessness, alongside a long-term programme of decommodified housing provision. But the IEA (among many other such institutions, economists, politicians and journalists) conjures up the threat of economic and housing disaster if any such regulation is imposed, arguing that landlords would not generate enough revenue to maintain their properties, leading to slum conditions, and/or simply withdraw them from the rental market. Unfortunately, just hearing the words 'rent control' is deeply unsettling to people who believe in so-called free and competitive markets, in private property rights and in the logic of trickle-down. The vast majority of economists, even some on the left like Paul Krugman (who famously trashed rent control in a *New York Times* column in 2000), are trained to think in the neoclassical way, striving for equilibrium through supply and demand. This perspective has become so hegemonic, so orthodox, that it did not take long for neoclassical economists to frame the fallout from the 2008 financial crisis as a time of 'recovery', rather than a time for any major structural/institutional reform or transformation (Mirowski 2013). But as the activists of Living Rent have been trying to point out, there is a world of difference between a home-owning economist working in a well-funded think tank or prestigious university receiving funding to research rent control via statistical analysis and a very low-income tenant struggling to pay rent for substandard accommodation whilst feeding and clothing their children.

As Sayer (2014) demonstrates, one of the most dangerous and pervasive myths of all is that the largely unearned income of the rich—which often includes owners of property and land—is actually being earned and is only fair given their 'hard work' and talent as 'wealth creators'. Unearned income derives from control of an already existing asset, such as land, buildings, technology or money, that others lack but need or want and who can therefore be charged for its use. It is the production of ignorance that stops so many from seeing that the wealthiest in society extract wealth from the economy without creating anything in return. As Living Rent recognises, the future achievements of movements for housing justice in the UK depend on their ability to organise and agitate for major changes in a demanding and hostile context of intentionally produced ignorance. Free market think tanks function as foot soldiers of, and mouthpieces for, the neoliberal creed. When that creed is under scrutiny or being challenged directly, think tanks scramble to produce 'evidence' (in the form of authoritative reports distilled into accessible sound bites) defending the free market and attacking state intervention in economic affairs. In this IEA case, structural causes of the housing crisis were swept out of sight by appeals to 'sound economics' and by recourse to the easily decipherable logics of supply and demand. Reports such as *The Flaws in Rent Ceilings* serve to divert attention away from the harsh realities of the housing crisis which lay bare the disastrous consequences of over three decades of neoliberal housing policies. Elsewhere (Slater 2014, 2016), I have written about how the influence of free market think tanks of the right is hard to avoid in any assessment of how the contemporary neoliberal state is aided and augmented. Their glossy and authoritative publications, their fast channels of access to authority and opinion-makers, their speechwriters and backroom 'researchers' have successfully deflected attention away from the causes of the increasingly miserable realities endured by people living at the bottom of the social and spatial order. Their contributions to the debate over housing have been instrumental in reframing a crisis of housing affordability as a crisis of housing supply (see also Gurran and Phibbs 2015, for the same dynamics in the Australian context). As the tactics and campaigning of Living Rent show, the intentional production of ignorance about the housing crisis is the context in which movements for housing justice in the UK now have to operate.

References

Anas, A. (1997). Rent Control with Matching Economies: A Model of European Housing Market Regulation. *Journal of Real Estate Finance and Economics, 15*(1), 111–137.

Baldwin, J. (1972). *No Name in the Street*. New York: Doubleday.

Ball, P. (2014, November 19). Gentrification Is a Natural Evolution. *The Guardian*. http://www.theguardian.com/commentisfree/2014/nov/19/gentrification-evolution-cities-brixton-battersea.

Bone, J., & O'Reilly, K. (2010). No Place Called Home: The Causes and Social Consequences of the UK Housing "Bubble". *British Journal of Sociology, 61*(2), 231–255.

Bourne, R. (2014). *The Flaws in Rent Ceilings* (IEA Discussion Paper No. 55). London: Institute of Economic Affairs. http://www.iea.org.uk/publications/research/the-flaws-in-rent-ceilings.

Buchanan, M., & Woodcock, S. (2016). Councils Spent £3.5bn on Temporary Housing in Last Five Years. *BBC News*. http://www.bbc.co.uk/news/uk-38016728.

Carlyon, T. (2013). *Food for Thought: Applying House Price Inflation to Grocery Prices*. Shelter Research Report. http://england.shelter.org.uk/professional_resources/policy_and_research/policy_library/policy_library_folder/food_for_thought.

Collinson, P. (2017, December 8). Four in 10 Right-to-Buy Homes Are Now Owned by Private Landlords. *The Guardian*.

Engels, F. (1845). *The Condition of the Working Class in England*. London: Penguin Books.

Gauldie, E. (1976). The Middle Class and the Working-Class Housing in the Nineteen Century. In A. MacLaren (Ed.), *Social Class in Scotland: Past and Present* (pp. 12–35). Edinburgh: John Donald Publishers Ltd.

Gurran, N., & Phibbs, P. (2015). Are Governments Really Interested in Fixing the Housing Problem? Policy Capture and Busy Work in Australia. *Housing Studies, 30*(5), 711–729.

Hamnett, C. (2010). Moving the Poor Out of Central London? The Implications of the Coalition Government 2010 Cuts to Housing Benefits. *Environment and Planning A, 42*(12), 2809–2819.

Hodkinson, S. (2012). The New Urban Enclosures. *CITY, 16*(5), 500–518.

Kutty, N. (1996). The Impact of Rent Control on Housing Maintenance: A Dynamic Analysis Incorporating European and North American Rent Regulations. *Housing Studies, 11*(1), 69–88.

Lansley, S., & Mack, J. (2015). *Breadline Britain: The Rise of Mass Poverty.* London: One World.

Leitner, H., Peck, J., & Sheppard, E. (2007). *Contesting Neoliberalism: Urban Frontiers.* New York: Guilford Press.

Madden, D., & Marcuse, P. (2016). *In Defense of Housing.* London: Verso Books.

Maloney, G. (2015). *A Living Rent for Scotland's Private Tenants.* http://www.allofusfirst.org/library/a-living-rent-for-scotlands-private-tenants/.

McCrone, D., & Elliot, B. (1989). *Property and Power in a City: The Sociological Significance of Landlordism.* Basingstoke: Macmillan.

McKee, K. (2016). Social Housing and the 'New Localism': A Strategy of Governance for Austere Times. In M. Bevir (Ed.), *Governmentality After Neoliberalism.* London: Routledge.

Meek, J. (2014). *Private Island: Why Britain Now Belongs to Someone Else.* London: Verso Books.

Minton, A. (2017). *Big Capital: Who Is London for?* London: Penguin Books.

Mirowski, P. (2013). *Never Let a Serious Crisis Go to Waste: How Neoliberalism Survived the Economic Meltdown.* London: Verso Books.

Olsen, E. O. (1988). What Do Economists Know About the Effect of Rent Control on Housing Maintenance. *Journal of Real Estate Finance and Economics, 1*(3), 295–307.

Pennington, J. (2015). *Analysis: Who Will Be Hit if the Benefit Cap Is Lowered?* Shelter Policy Briefing. https://england.shelter.org.uk/professional_resources/policy_and_research/policy_library/policy_library_folder/analysis_who_will_be_hit_if_the_benefit_cap_is_lowered.

Proctor, R., & Schiebinger, L. (Eds.). (2008). *Agnotology: The Making and Unmaking of Ignorance.* Stanford, CA: Stanford University Press.

Rodger, R. (1989). Crisis and Confrontation in Scottish Housing 1880–1914. In R. Rodger (Ed.), *Scottish Housing in the Twentieth Century* (pp. 25–53). Leicester: Leicester University Press.

Rolnik, R. (2013). Late Neoliberalism: The Financialization of Homeownership and Housing Rights. *International Journal of Urban and Regional Research, 37*(3), 1058–1066.

Savills. (2017). *London's Future Homes and Workplaces – The Next Five Years*. London: Savills World Research.

Sayer, A. (2014). *Why We Can't Afford the Rich*. Bristol: Policy Press.

Shelter. (2014). *Safe and Decent Homes: Solutions for a Better Private Rented Sector*. http://england.shelter.org.uk/__data/assets/pdf_file/0003/1039530/FINAL_SAFE_AND_DECENT_HOMES_REPORT-_USE_FOR_LAUNCH.pdf. Accessed 6 June 2017.

Slater, T. (2014). The Myth of "Broken Britain": Welfare Reform and the Production of Ignorance. *Antipode, 46*(4), 948–969.

Slater, T. (2016). The Housing Crisis in Neoliberal Britain: Free Market Think Tanks and the Production of Ignorance. In S. Springer, K. Birch, & J. MacLeavy (Ed.), *The Routledge Handbook of Neoliberalism*. London: Routledge.

Stafford, D. C. (1976). The Final Economic Demise of the Private Landlord? *Social and Economic Administration, 10*(1), 3–14.

UK Parliament. (2016). *Housing and Planning Bill: Written Evidence Submitted by Crisis* https://www.publications.parliament.uk/pa/cm201516/cmpublic/housingplanning/memo/hpb04.htm. Accessed 6 June 2017.

Watt, P., & Minton, A. (2016). London's Housing Crisis and Its Activisms. *CITY, 20*(2), 204–221.

CHAPTER 4

Urban Food Activism in Athens: Recovering More Autonomous Forms of Social Reproduction

Inés Morales-Bernardos

Introduction

As the feminist philosopher of the autonomist Marxist tradition Silvia Federici witnessed after visiting various spaces in Athens that emerged between the mobilisations of 2008 and 2015: 'neither capital nor the state provide any means of reproduction-they exist only as repressive forces, so many have begun to pool their resources and create more collective forms of reproduction as the only guarantee of survival' (Federici and Sitrin 2016). During the last wave of neoliberalisation, the slashing of social protection, coupled with the destruction of waged labour and insufficient incomes, unfolded a crisis of social reproduction and care with important relevance in the southern European peripheries (Hadjimichalis 2014; Hadjimichalis and Hudson 2014; Lekakis and Kousis 2013; Zechner and Hansen 2015).

I. Morales-Bernardos (✉)
University of Córdoba, Córdoba, Spain

N. M. Yip et al. (eds.),
Contested Cities and Urban Activism, The Contemporary City,
https://doi.org/10.1007/978-981-13-1730-9_4

More visible across urban spaces, such as in the city of Athens, these difficulties in the reproduction of lives have triggered various mobilisations and forms of urban activism over the question of social reproduction. Food, water, housing, education, to name only a few, have been decisive sites of struggle. In doing so, activists have performed[1] particular repertories through building new collective spaces of social reproduction for those impoverished, excluded or unrecognised by the social protection of the state, as well as through special efforts to shape non-racist and inclusive spaces for those affected by the rise of nationalism and far-right activism (Koronaiou and Sakellariou 2013).

The particular relevance of building spaces for food provisioning, challenging the food scarcities that came out of the previous wave of neoliberalisation in 1990s Athens (Arampatzi 2012; Dalakoglou 2013a) and following the desire to 'take their lives back into their hands' (Various Activists, Athens, 2015), shaped a particular urban food activism. Urban gardens, social kitchens, food banks and farmers' markets, to name only a few spaces, have been set up, shaping a new food geography and a particular politics of caring and food provisioning, recovering more autonomous forms of social reproduction. The preservation of traditional seed varieties, food growing in the city, food expropriations in big food chains or cooking together have been some of the new collective activities performed in everyday life by food activists.

Thus, in this chapter, we aim to explore the forms of urban food activism that emerged and were performed between the 2008 and 2015 mobilisations in these peripheries following the desire to 'take their lives back into their hands', their repertories and internal dynamics (organisational forms, political concerns), and their relationship to their context (alliances with other forms of activism, relations with political parties, NGOs) through an analysis of food geography and the politics of caring and provisioning that they performed. Furthermore, we aim to explore the tensions that resulted from shifts in their relations with the state and how they proposed a heterogeneous 'politics of autonomy' (Castoriadis 1991; VVAAs 2011; Federici 2012; Newman 2011, 2016; Nasioka 2017; Gutierrez 2017), or the more institutional 'social and solidarity economies' (Rakopoulos 2014; Petropoulou 2014; Arampatzi 2017), differently after the 2012 elections.

[1] Following Henri Lefebvre's (2009) and Judith Butler's (2015) performative philosophy, we use a theatrical perspective in this chapter.

We argue that the existing conceptual and theoretical framework traditionally used to understand the dynamics and potential of urban activism, and, more specifically, food activism in European contexts needs to be reconsidered if we want to understand the particularities of urban food activism in Athens between the mobilisations of 2008 and 2015, their tendencies towards non-state-centric politics in organising social reproduction, and that, in agreement with other authors, they regularly adopt more invisible spatialities and temporalities (Lefebvre 2009; Carabanchaleando 2017; Gutierrez Aguilar 2017).

Following these assertions, the form of food activism observed has not only been dealing with the contradictions of corporate food systems, as have other forms of food activism performed in Athens and other European cities. Instead, these urban food activists are particularly and more widely engaged with the specific socio-contradictions of the capital and state organisation of social reproduction in these urban peripheries, and which activists consider responsible for food scarcities in the European peripheries—that is, the lack of time and space for caring and food provisioning.

The arguments raised in this chapter draw on one year of militant research within various spaces established after the December 2008 revolts and which functioned until the 2015 anti-austerity referendum and mobilisations. The participation in the various spaces, the cooking or growing of food, has been complemented by taking notes during discussions and assemblies, and with more than 50 semi-structured interviews with some of their participants.

The original empirical findings gathered during these activities in Athens and the literature review build the body of this chapter. In order to understand their politics and the forms through which they built their repertories, we have first considered their practices. We give important relevance to their readings and political philosophies with whom they identified themselves, selecting authors such as Silvia Federici whose reflections introduce this section. As various activists shared with us: 'Athenian activists are really conservative. We read and read, and we discuss in order to analyse each of our moves' (G. 2016).

Then, to understand the particularities of these forms of urban activism in Athens between the mobilisations of 2008 and 2015, we first introduce the method of analysis developed, and in particular, we try to approach the dynamics of social reproduction in the urban peripheries. Second, we analyse the particularities of urban food activism in

Athens, the food geography and their politics of caring and food provisioning. Third, we discuss the main tensions that have been shaping these spaces.

An Approach to Urban Food Activism from the Peripheries

The existing conceptual and theoretical framework traditionally used to understand the dynamics and potential of urban activism, and, more specifically, food activism in European contexts, needs to be reconsidered if we want to understand the particularities of urban food activism in Athens between the mobilisations of 2008 and 2015 and their more autonomous tendencies. As we argue in this chapter, and agreeing with Raquel Gutierrez Aguilar (2017), traditional categories focused on state-centric politics regularly render invisible the temporality and spatiality, and the repertories and political philosophies of urban activism in the peripheries.

Thus, the particularities of food activism in these peripheries that we argue need to be considered are: their urban dimension; their concerns in reorganising social reproduction, caregiving and food provisioning beyond state and capital; the specific dynamics of social reproduction in the peripheries, and more precisely, in Athens (that we approach in the next section); and finally, the particular temporalities and spatialities of their repertories, which are regularly embedded in the activities necessary for the reproduction of lives.

Thus, we first need to bear in mind the relevance of the urban dimension of these forms of food activism. With this consideration, and by reading Margit Mayer's reviews on Manuel Castells' work on urban mobilisations in Western countries during the 1960s (2006) together with her more recent works on uprising and urban activism (2013, 2016), we must consider not only the centrality of the 'reproductive sphere' and issues of 'collective consumption' in their politics and repertories in order to understand contemporary urban mobilisations, but also, as they argued, the particular tensions that activists have in relation to the state.

Considering that caregiving and food provision provided by these forms of activism are part of the 'reproductive sphere', that is activities involved in social reproduction, we need to clarify our understanding of this term. Used in Marxist thought and, more recently, by Marxist and socialist feminists since the 1970s, social reproduction has acquired a wide array of meanings and has been put to many different uses (Katsarova 2015). However, in this chapter, we consider social reproduction as both the material means of subsistence and survival, as well as what sociologist, Raquel Gutierrez, in an interview with Silvia Federici described as: 'the set of activities and cyclic processes needed for the reproduction of life as a whole' (Gutierrez 2015, p. 1).

Reading theoretical frameworks about movements around the question of social reproduction and their relation to the state (Nadasen 2012; Armstrong 2011, 2015; Katsarova 2015), we can understand the activists' perception of the state as being repressive or as a protector, and their repertories, such as demanding and participating in decision-making or instead, adopting more antagonistic relations with state power, have been importantly shaped by particular historical interventions of the state in the dynamics of social reproduction. We try to approach these perceptions in the case of Athens in the next section. Contributing to this tension, particularly in the urban context, Margit Mayer referred to the different strategic positions activists historically took, depending on how they could benefit from social protection or how they were either able, or willing, to open spaces of negotiation and participation in local governance (Mayer 2013).

Consideration for forms of food activism we are exploring within the more unified and recognised existing categories, offered through the work of Holt-Gimenez and Shattuck (2011), has been limited. Holt-Gimenez and Shattuck's categories, food sovereignty and food justice movements (2011), are focused on the dynamics of food systems (production, reproduction and consumption) without allowing for activities such as caregiving, and the wider implication of food activism in social reproduction. Furthermore, these categories do not consider the different perceptions and relations of activists with the state and its role in organising social reproduction. Their categories are drawn from the 'the potential for food movements to bring about substantive changes to the current global food system', and the capacity of activists to build local and democratic food systems in order to challenge the 'right to food' or the 'human right to food'. However, they do not consider the different perceptions and roles that local states have in social reproduction, either as being repressive or as protector, nor the antagonistic relationship or opportunities of activists in opening spaces of negotiation to 'bring about changes into the local and global food systems'.

This last argument coincides with those of Armstrong (2011) and Katsarova (2015): the ambiguous relationship with the state in Western countries that movements around the question of social reproduction regularly adopt. As these authors pointed out, while these movements 'criticized the disciplinary and racializing practices of the state, they continued to levy their social demands on the state organized around forcing the state to give adequate benefits, while also fighting against the "dehumanizing and surveillance-based components of welfare"' (Armstrong 2011).

Armstrong and Katsarova perceptions go along with the ambiguous relationship some activists have with municipal authorities in cities when they benefit from social protection and participate in the decision-making. As Margit Mayer reflected (2013, p. 11), the particular activism which adopts a cooperative relationship with local states is performed by the 'creative classes' (Markusen 2006; Colomb 2012). In the last wave of neoliberalisation, they had been involved in the development of 'creative city politics' (Mayer 2013, p. 11) and who, through the city branding strategy, allowed cities to enter into the global competition of 'smart cities'. Within food activism, activists who promote urban agriculture and healthy local food systems in particular, coinciding with other authors (Darly and McClintock 2017), are contributing to the new wave of neoliberalisation in cities as they are improving the city image. The urban food activists within this trend would be more interested in the well-being and ecological sustainability of cities (see Milan Urban Policy Pact).

A last consideration needed in order to understand the particularities of the urban food activism approached in this chapter is their invisible and controversial tendencies. Some scholars refer to them as 'movements of the urban outcast' (Mayer 2013, p. 12). Others have been referring to them as the 'cultural effects of the protest movements' (della Porta 2017). While still other authors consider them as 'weak', 'spontaneous' and 'informal' in comparison with Western countries and the categories of 'strong', 'formal' and 'organised', they assign them instead (Leontidou 2010, 2014; Petropoulou 2014).

In this chapter, we argue that urban food activism from the peripheries is particularly invisible, not only to some scholars, but also to other activists, due first to their specific temporality. The caregiving and food provisioning these activists are performing are embedded in the everyday activities needed for the reproduction of life, whose lack of recognition and visibility feminist activists and scholars have been denouncing (Federici 2012; see Viewpoint 2015; Bhattacharya 2017). Second, due to their spatiality. The repression they regularly face by their local states and activist organisations, as they consider social transformation and organisation of social reproduction beyond state-centric politics or the mediation of any other institution (NGOs, political parties), force or induce them to adopt a particular spatiality. Despite some other forms of urban activism that are able to perform activities and repertories in public spaces more freely, these forms of activism correspond to those referred to by James Scott as 'weapons of the weak' (Scott 1985) and by Judith Butler as the precarious ones who are not normally 'recognised' by others (Butler 2015). Furthermore, these forms of activism in the peripheries could be consistent with anarchist philosophies. These forms of activism are performed in public spaces, but they do not aim to be recognised by the state (Graeber 2002; Newman 2007, 2016; Springer et al. 2012).

As mentioned, this particular form of food activism does not demand the right to food or the human right to food by building spaces of negotiation with the local state. They do not demand more democratic food systems or better infrastructure for collective consumption. They instead, corresponding with Saul Newman's ideas around autonomous spaces (2011, p. 345), 'create in their practices, discourses and modes of action, new political, social and economic spaces, new imaginaries. […] What shapes this alternative political space is, I would argue, the idea of autonomy'. More precisely, they build a particular politics of autonomy, which, following Raquel Gutierrez arguments (2017, p. 66), inform the repertories and philosophies of what she designated as 'politics in feminine': non-statist, transformative and subversive politics that build collective, inclusive, spaces for the reproduction of lives.

These more autonomous forms of activism, as the geographer and philosopher Henri Lefebvre already pointed out (Lefebvre 2009) in addition to other authors (Gutierrez 2017; VVAAs 2011), normally emerge in the peripheries of capitalist social organisation, where states have less presence, or where activists, through some specific events as John Holloway also pointed out, can open cracks in capitalism (2010).

Urban Activism in Athens: Antecedents

As an approximation in understanding the contemporary conflicts and urban mobilisations in these peripheries, we need to approach the historical social reproduction dynamics in Athens. Athens can be understood as peripheral to the economies of Western European countries, semi-peripheral to the world economy and with specific internal dynamics of exploitation and oppression, as well caregiving and food provisioning.

The dynamics of social reproduction, activities and processes in the reproduction of life, and provisioning and caring, as some authors have argued, have historically been confronted with the intervention and organisation of state and capital (Federici 2004; see Viewpoint 2015; Fraser 2017). As Nancy Fraser stated, the intervention of state and capital turned the more traditional autonomous forms of reproduction into the commodified and privatised ones we find today (Fraser 2017, p. 25), provoking various crises, impossibilities or difficulties in social reproduction.

Reading different historical works, and looking today towards other peripheral cities where more autonomous forms of reproduction remain or have been recovered (VVAAs 2011; Gutierrez 2017), autonomous forms of reproduction would ideally underpin the provision, particularly food provision, of communal resources such as land, seeds and water from the surrounding countryside, while caring would be performed through mutual, horizontal relations in heterogeneous forms of communities, with ideally no gender division, no labour division and no oppressive relationship with nature.

According to Mohandesi and Teitelman (2017, p. 40), the shift between the autonomous and the commodified traditionally happened through several related, though distinct, historical pressures that underpin social reproduction in waged labour. As they also pointed out, communities that were not dependent exclusively on capitalist relations were violently persecuted, and their communal resources either expropriated or destroyed, not only in the countryside, but also in the city (city pigs, crops, vegetable gardens). Violent persecutions have also historically provoked political contestation (Federici 2004; Mohandesi and Teitelman 2017, p. 40; Fraser 2017, p. 25; McMichael 2009).

In the city of Athens, the first major shift and pressures towards a dependency on waged labour were historically documented by various authors in the 1950s, after World War II and the Civil War, Emfylios (Leontidou 1990, 2010; Makrygianni and Tsavdaroglou 2011, p. 29; Kritidis 2014, p. 63). Carried out by post-war authorities,[2] the Marshall Plan's aim was the modernisation of the 'underdeveloped countries'; it pushed for a vast urbanisation of Athens and soft industrialisation. Under the Marshall Plan, the Green Revolution brought about 'agricultural development' (Shaw 1969, p. 383) and the industrialisation of various traditional forms of agriculture in order to 'reorganize production towards higher value products' (Shaw 1969, p. 35).

With no clear agreement between these authors as to whether the process of industrialisation was the main driver of the first major shift, but consistent with some of their arguments (Makrygianni and Tsavdaroglou 2011, p. 29; Kritidis 2014, p. 63), the mechanisms of pressure that allowed the underpinning of waged labour in Athens were first, during the post-war era, achieved through impoverishment and political repression of the countryside intended to provoke the abandonment of lands and a massive exodus of the rural population to the city of Athens; and second, the demolition or legalisation of self-constructed neighbourhoods that these newcomers built in the suburbs of the city (Leontidou 2010; Makrygianni and Tsavdaroglou 2011, p. 29) starting in the 1960s under the dictatorship of the Colonels. According to these authors, the building of the afthereta, new neighbourhoods in squatted land in the outskirts of the city and where the populations were organising autonomously, can be considered the first major forms of urban activism in Athens, challenging both the repressive-protector role of the state and the forms of organisation of social reproduction in the city.

During the Emfylios, provision started to be organised partially by capital, shaping what could be understood as hybrid forms of reproduction. While capital provided waged labour and commodities such as food resulting from the industrialisation of agriculture, more autonomous forms of reproduction (traditional agriculture and resources coming from the surrounding countryside, sharing of incomes [Leontidou 2010], as well as caring within families and wider communities in more horizontal relationships) were still importantly sustaining social reproduction.

[2]The Greek Government supported by the American Mission for Aid to Greece, see: http://library.cqpress.com/cqresearcher/document.php?id=cqresrre1949020900e.

Later in the 1970s, as some scholars have argued (Federici 2004; Fraser 2017), autonomous forms of reproduction suffered major pressures. The change in the global dynamics of capital accumulation provoked a major global shift and tendency towards pushing social reproduction into waged labour. Importantly, it was at this point that the invisible reproductive work of caring and provisioning began to be commodified and privatised.

The major global pressures in this new period were, globally, the mechanism of debt imposed by 'financialised capitalism' (Fraser 2017, p. 32), and, more locally, in those countries with welfare states, the social protection of the local states (Armstrong 2011, 2015).

In Greece, in the city of Athens, this new order has been conducted through various processes starting in the 1970s, continuing into the metapolitefsi (post-dictatorship era) until 2015 with the acceptance of the new order ruled by the Troika. Various global financial bodies and a newly democratic welfare state have intervened in the organisation of social reproduction. In this new organisation of social reproduction, financial capital has ideally aimed to provide full employment as well as commodities with better organisation of production. Two main financial bodies have been reorganising social reproduction in this period: first, since the 1980s, the European Union, and later beginning in 2010, the Troika—the European Union together with the European Central Bank and the International Monetary Fund. In the case of agriculture, via different legislation at a European and global level, they aimed to concentrate the control of the organisation of food production and provision within a few food chain companies. Social protection—health care and education in the case of Greece—was provided by the local welfare state, while the European Union and United Nations regulated and controlled their implementation (Baker et al. 1997).

According to Athina Arampatzi, the major difficulties of this period in Athens emerged after the 1990s with a new round of neoliberalisation (Arampatzi 2012, 2017) destroying waged labour, diminishing salaries and privatising space, education and health care. The climax of this round of pressures unfolded in 2010 with the destruction of wage labour, new measures of privatisation and the slashing of social protection by the Troika (Hadjimichalis 2014).

Autonomous forms of reproduction struggled with major pressures during the metapolitefsi. The social reproduction that importantly remained in Athens under forms of waged labour and the sharing of incomes among family members, as well as between wider communities, was threatened by insufficient salaries, the lack of waged labour and the destruction of pensions. Moreover, the provisioning of food from large food chains and food industry intermediaries was threatened by increased taxation and increased regulation on the more direct forms of provision. More autonomous forms, however, such as weekly traditional farmers' markets or small farming in the countryside, provided, through direct relations and different forms of trading, fresh vegetables to every neighbourhood in Athens, at lower prices. The work–life balance was also distorted due to the flexibilisation of labour and the diminished capacity (insufficient quality of time and space) for caregiving to families and wider communities. A lack of time for cooking, eating together and building social bonds and affections has made it difficult to provide care.

Various urban mobilisations occurred around the question of social reproduction during the metapolitefsi that preceded the urban food activism we analyse in the next section. Student occupations[3] and riots on the streets had already begun in the 1980s, intensifying in the 1990s with temporal occupations of universities and high schools (Vradis and Dalakoglou 2011, p. 91; Kritidis 2014, p. 63). During the 1990s, mobilisations also occurred against the privatisation of public space, health care and education, demanding better social protection from the state (Arampatzi 2012, p. 2598; 2017). These mobilisations were violently repressed by police, ending occasionally with the assassination of activists and students (Vradis and Dalakoglou 2011, p. 91; Dalakoglou 2012, 2013a; Kritidis 2014, p. 63).

In their main repertories, apart from temporal occupations, some social spaces were set up (squats, social centres, stekia and neighbourhood assemblies), organising more autonomous provisioning and caring. These spaces were specifically located in the central neighbourhood, where the occupation of the universities ended with the dictatorship, and where historically activists, students and migrants settled during the metapolitefsi. In these spaces, they organised alternative education

[3]The occupations of the Athens Polytechnic and the Business School in November 1973 caused the end of the Colonels' dictatorship, the *Junta*. These occupations, as Kritidis referred to them, 'marked the beginning of a new era'. 'The action was a turning point, because it was the first time in Greece that public buildings were occupied and transformed into centers of social and political mass protest. Eventually, the action caused the end of the military dictatorship' (Kritidis 2014, p. 66).

for migrants, but also the provisioning of food via direct and solidary forms of trading between consumers and farmers, urban gardening and the preservation of seeds. The repertories were inspired by other movements: memories of the Spanish Social Revolution of the 1930s; or more contemporary ones such as the Zapatistas in Mexico, the 'Piqueteros, asambleístas' and 'fábricas recuperadas' of the 2001 Argentinian uprising; and other European urban autonomous movements. Nevertheless, the mobilisations of 2008 and 2015 acquired a more particular spatial dimension after the assassination of Alexis Grigoropoulos by a police officer, provoking insurrections in December 2008.

Urban Mobilisations and Food Activism in Times of Food Scarcities in Athens at the End of the Metapolitefsi

Preceded by the various mobilisations recounted in the previous section against privatisations of space, health and education, and demanding better infrastructure from the state, two major uprisings and massive occupations transformed the everyday politics and repertories of urban activism in Athens. As we explore in the next sections, within this period they reorganised social reproduction in more autonomous and collective forms, spreading the spaces that had been set up previously for education, food provisioning, health care and housing.

The first uprising, the 2008 December revolts, occurred after the assassination of Alexis Grigoropoulos by a police officer in the historic political neighbourhood of the city centre (Vradis and Dalakoglou 2011; Nasioka 2017), while the second, the 'Movement of the Squares' in 2011, occurred after the intervention of the Troika (Dalakoglou 2012; Nasioka 2014; Stavrides 2014). This last one followed a wave of mobilisations and occupations in other major cities around the world (Mayer et al. 2016; della Porta 2017; O'Brien 2017), and especially in other southern European cities where the welfare states were implementing austerity measures, privatisations and slashing social protections following Troika intervention (Hadjimichalis 2014).

These two major uprisings spread the desire to, as activists refer to it, 'take our lives in our own hands' (A, G, K, E, P, activist, Athens, 2015), sparking waves of occupations and the establishment of new social centres, squats and neighbourhood assemblies all over the city. Activists transformed some of the spaces into urban gardens, social kitchens and health clinics, to name a few. Building spaces for food provisioning, they shaped a particular food activism that was gradually embedded into the everyday activities needed for the reproduction of life.

Movements with no demands resulted, and, despite the relevance of the uprisings, the politics of everyday life gained centrality. As some other authors highlighted after visiting Athens (Agamben 2015; Federici and Sitrini 2016), the more authoritarian and repressive role of the state in particular contributed to the shape of a particular politics of autonomy and the antagonistic features of these forms of activism. Thus, willing to de-commodify social reproduction and organise it without the intervention of local authorities, NGOs or political parties, they built spaces specifically through prefigurative politics, assemblies and horizontal decision-making, shaping the particular politics of caregiving as well. As in previous years, but more intensified, these antagonistic activists faced repression not only from their local states, but from other activist organisations, far-right activists and left-wing activists as well (Dalakoglou 2012; Markantonatou 2015).

Predominantly, after the second wave of occupations and the implementation of new austerity measures, the food scarcities and the difficulty of farmers in the countryside to continue their activities led to various other mobilisations, diversifying the repertories of food activism as well. Activists, with the support of farmers, set up in city street markets, food cooperatives, food banks and groceries stores where they were able to provide food, fresh food or cooked meals in different neighbourhoods of the city.

The lack of employment in the city also led to the formation of working cooperatives or collectives in the form of grocery stores in the city and farming ventures in the countryside. Although they were supported by activists and took more direct forms of provisioning, such as the distribution of the kalazi, or basket, they adopted a more entrepreneurial dimension. These forms of food activism were closer to the dynamics of the construction of 'sustainable healthy local food systems' that local authorities, with the support of various European programs and the Food and Agricultural Organization (FAO), had been promoting in various cities (see Urban act; see Milan Pact).

The rise of the left-wing party Syriza in 2012 and the creation of a new NGO (Solidarity for All) willing to intervene in the organisation of social reproduction that activists were performing importantly reshaped the politics of everyday life, the forms of organisation and their repertories, the spatiality and temporality. These events led to cooperation with these organisations on some of their activities or their delegation, to a demand for infrastructure of the local authorities and to re-commodify the more autonomous forms of reproduction through the new paradigm of 'social and solidarity economies' (Rakopoulos 2014; Arampatzi 2017).

The rise of nationalism and the slashing of social protection also brought about the rise of far-right activism and the consolidation of the neo-Nazi political party Golden Dawn (Vasilopoulou and Halikiopoulou 2015). This included attacks on migrants and political activists, in addition to the establishment of social kitchens and food banks—for the impoverished Greeks exclusively (Koronaiou and Sakellariou 2013).

Following the desire to take their lives back into their own hands, with continuous tensions between their more antagonistic or cooperative relations with state, these various tendencies have been building a new food geography in Athens, but one with a particular anti-fascist shape (anti-racist and horizontal). They performed occasional convergences, causing major tensions and shifts between the politics of autonomy in the first wave of mobilisations, and the social and solidarity economies of the second one. These tension are the subject of analysis in the section "Food Geography in Athens: The Invisible Spatiality and Temporality of Urban Food Activism".

Food Geography in Athens: The Invisible Spatiality and Temporality of Urban Food Activism

'A completely different time and space from what we have experienced before was created. We felt we could intervene on the reality in a more direct form. We felt that we could solve the problems of the city' (K., an activist who participated in the 2008 and 2011 uprisings, Athens, 2015).

Coinciding with the memories K. shared with us, urban food activists shaped a particular food geography by building more autonomous forms of reproduction. Their controversial repertories, based mainly on the squatting of public and privately owned spaces, and the lack of recognition by local authorities and other activist organisations, avoiding intermediaries (NGOs, political parties and the food industry) while performing their activities, particularly contributed to the invisible dimension of this geography as well.

Stemming from their desire to 'take their lives back into their hands', the collective self-organisation dimension importantly shaped the new food geography. By reorganising collectively beyond institutional forms of organisation, they were either hosted in already existing collective spaces or occupied new ones: squatted or rented. And when in need of new spaces, in most cases they squatted publicly owned buildings, abandoned or privatised during the dismantling of the welfare system.

As in previous years, these collective spaces took the form of squats, social centres, stekia and neighbourhood assemblies. More specifically, hosted in these spaces, either separately or as a collective identity, were urban gardens, social kitchens, self-organised food banks, food cooperatives, grocery stores or farmers' markets set up so that farmers, activists and consumers with different identities could gather and negotiate their needs.

These collective spaces were also shaped differently internally due to the need to organise and host concrete activities for food provisioning and caring—not only for the more technical needs, but also for collective decision-making. Within the social centres and squats as well as in food cooperatives, they have built separate spaces for the preservation of seeds, for their storage, for making seedlings and growing vegetables. They also built spaces differently for the storage of food collected for food banks, the preparation of meals and the possibility of eating together in social kitchens. Some spaces were specifically created in order to make decisions collectively, host their internal assemblies and organise public events with other collectives.

The food geography also took on a decentralised dimension in order to answer the specific needs of different city neighbourhoods. The spaces we referred to in the previous sections were built in the central neighbourhood, but also in various neighbourhoods, and from them, to other cities and the countryside, mainly in the form of farming ventures.

Their forms of communication also gave a spatial dimension to their repertories: mainly posters fixed on the walls of their neighbourhoods, as well as social media to communicate broadly with other neighbourhoods and cities. For their internal communication, they used mostly telephones and email.

As referred to above, a particular temporality was given to the forms of food activism in Athens. Given the changing politics and repertories of these activists—collectively reorganising social reproduction and answering to the changing everyday activities needed for the reproduction of life—they adopted more widely the prefigurative politics of the here and now, which also contributed to a permanent dimension in their repertories. Although they took the street in protest at first, afterwards they turned their occupations into collective spaces, willing to answer everyday food and caring needs.

Nevertheless, some of the spaces retained a temporary dimension—those used for major coordination with other collectives and activists. In this way, they used gardens or squares to organise festivals, gatherings or occasional public assemblies. There was also weekly food distribution from some farmers in kalazis or baskets; farmers' markets organised once or twice a month in many neighbourhoods; and cooking in the streets.

Caring While Growing Vegetables and Cooking

In building this geography, urban food activists gave an important dimension to caregiving. By taking part in these spaces and reading the public statements of various spaces (social kitchens, urban gardens, food cooperatives and so on), we understood they were willing to answer everyday food needs through self-organisation, with non-sexist, non-hierarchical, non-racist relations, multi-class spaces, but also particularly anti-fascist everyday relations and repertories, building a particular politics of care.

As observed while cooking or growing food, prefigurative politics gave a particular repertoire (assembly, horizontality) to the better self-organisation and negotiation of the individual and collective needs of the various identities that converged in the same spaces: farmers, consumers, migrants, unemployed, workers, precarious, de-classed, students, men, women, activists, receivers and especially, as observed, those non-Greek, non-Europeans rejected by far-right activists and the social protection of the state, as well as those with psychological problems and drug or alcohol addictions, who dramatically increased during this period.

Throughout the opening of a collective dimension to the provision of food, multi-gender assemblies or councils were the main body of organisation and decision-making. Although some collectives decided to organise themselves through more structured or hierarchical forms of organisation, taking the form of associations, most spaces were organised horizontally, assuming collective responsibilities in the provisioning of food or decision-making and trying to build their internal dynamics through cooperative, mutual aid or solidary relations, avoiding the division of labour and work relations. This particularly opened tensions for those willing to build working cooperatives or willing to accept the intervention of NGOs and political parties or delegation to them.

As observed and, in comparison with other social and political spaces, more women actively participated and made decisions in these spaces, mediating assemblies, taking minutes or cooking. Challenging the normative gender divisions of reproductive work as well, men were importantly present and involved in cooking and caring.

While the principles of non-racist relations notably shaped the ethics and politics of these spaces, they were avoiding having relations with farmers, consumers and other activists that could reproduce authoritarian and violent relationships. After some incidents on various farms, and the increasing employment of migrants in farming labour, they specifically gave importance to the working conditions of migrants.

As previously mentioned, the rise of far-right activists who set up social kitchens and food banks for impoverished Greeks triggered various conflicts between left-wing and far-right activists. These tensions finally led to particular non-fascist everyday relations within these spaces, and instead of cooking for Greeks, they cooked for everyone. In this way, they contributed with more repertories beyond the regular more confrontational ones that anti-fascists activists performed normally in Athens.

New identities of receiver and activist-supporter were shaped. The solidary and philanthropic features of these spaces were negotiated, as were the paternalistic relations that occasionally resulted. These particular relations brought bigger discussions and tensions to the spaces due to their own understandings and practices of solidarity. In particular, interested in not reproducing the exploitative relations of capital, they attempted to build direct relationships between farmers and consumers, avoiding the intermediaries of the corporate food system.

One of their major concerns was the class desegregation of the spaces. Influenced by the activists' previous participation in unions or workers political parties, they observed the working relations within their spaces. Some of them avoided any working relationship, without a division of labour, establishing a voluntary–militant work scheme with no bosses and no workers, while others tried to overcome exploitative relationships with new formulas as work cooperatives.

The major divisions concerning class identity were due to the monetised or non-monetised character of the spaces. The social kitchens of food banks were among the more inclusive considering the particular needs of each of the participants. Food cooperatives or grocery stores providing organic food were more segregated, attracting mainly those with regular waged labour and the middle classes and almost no non-Europeans among them.

The various political identities also caused major tensions and divisions. The highly politicised spaces and, as one of the interviewed referred, the 'conservative character and different fantasies' of activists in Athens divided the spaces between anarchists and lefties. The different approaches to relations with the state caused major tensions after 2012, as we analyse in the section 'Between the Autonomous and the Hybrid Forms of Social Reproduction'.

Politics of Food Provisioning

In building their particular politics of food provisioning, their major tension was the relationship between capital and nature. Navigating between more autonomous forms of reproduction and more hybrid ones, they were building their own principles in order to collectively provide food: non-profit, fair prices, produced at the local level, without intermediaries, preferably without employees on small farms and through the non-use of chemicals.

The provision of food at quantities for a major portion of the population through non-monetised circuits can be considered their main challenge and division in the politics of food provisioning, and they were approaching it through various politics of autonomy, or alternative economies, within a paradigm of social and solidarity economies.

In their diverse politics of autonomy, unwilling to establish relationships with capital or the state, they tried to rebuild more autonomous forms of social reproduction through mutual aid and solidary relations, preferably supporting the self-provision of food through voluntary or militant work with no separation of labour and with no separate identities between farmer-consumer and supporter-receiver. With the multiple possibilities of their organisation, they could start provisioning with the preservation and replication of their own seeds, growing their own food or provisioning through different forms of relationships that avoided monetised value. When they had considered the monetised value of food, they exchanged food through contributions, having no fixed prices, or by exchanging other material needs. Time banks were also used in these relationships. Some of these spaces were named 'self-defense structures', 'anti-capitalist commons', or 'self-organised spaces'. Their main obstacle was the impossibilities of building food autonomy in the city of Athens thanks to the lack of land and resources to grow food in the city, as well their constraining need for waged labour to supply other material needs.

The hybrid forms of food provisioning considered were the opening spaces of negotiation with NGOs and political parties, and the provision of food through monetised forms of exchange. They aimed to build provisioning through solidarity relations: run through waged labour and workers, taking the form of working cooperatives sometimes supported by activists, and voluntary or militant work. The paradigm that frames these forms of provision were the social and solidarity economies. Their main constraint was their dependence on resources dispensed by NGOs, political parties or European programs.

Although they had always been secondary to the provision of a large enough quantity of food and the provision of food at fair prices, the quality of food and the use of non-chemicals were also considered. In this way, they considered different farming techniques such as organic farming, permaculture or agroecology. Most of the farmers practising these techniques took part in what was considered a back to the land movement. Young, educated and unemployed urbanites started farming ventures in the countryside; they provided food through a weekly kalazi (basket), delivered to some common spaces and door by door in the city of Athens.

The local dimension also brought tensions. The special interest in building local food systems that supported the Greek economy and Greek farmers often converged with far-right activists sharing an interest in supporting Greeks. Another major tension in the organisation has been the lack of responsibility taken by the farmers for the collective organisation, delegation of distribution and participation in assemblies to consumers-activists or to specific workers.

Between the Autonomous and the Hybrid Forms of Social Reproduction

Discussing and agreeing with various activists, one of the major tensions that shaped the food geography in Athens of the period analysed was the shift between the activists' proposals for more autonomous forms of reproduction and those underpinned by the new, more institutional paradigm of social and solidarity economies supported by the NGOs and political parties. As mentioned before, this major shift occurred after the rise of the left-wing party Syriza in 2012 and the establishment of a new NGO (Solidarity for All) willing to intervene in the organisation of more autonomous social reproduction that activists were performing. Their repertories of self-organisation turned into delegation, and their desire to 'take their lives back into their hands' turned into demanding infrastructure of the local authorities. The antagonistic character of some of these forms of activism turned cooperative with the local authorities. As many of the activists noted, the cooperation with the NGO, who for some interfered in the organisation and coordination of the various spaces, caused tensions and divisions among them. While the welfare systems were being dismantled, framing these spaces as solidarity economies offered local authorities the innovative forms activists were building to solve problems of unemployment, health care and food provision, among others, and replaced the more autonomous forms of reproduction with more temporal and delegative ones. With their more structured and hierarchical forms of organisation, some of these spaces either ended their activities or transformed them into more entrepreneurial ones: from self-organised farmers' markets to grocery stores.

The professionalisation and entrepreneurial tendency, along with the divisions of the spaces, became clearer after the 2015 elections when the same anti-austerity party was elected but accepted a new bailout and the implementation of austerity measures. The increasing economic and social pressures limited the participation and the repertories of many activists leading to a considerable demobilisation.

In times of increased neoliberalisation, another major difficulty these activists were facing while performing social and solidarity economics and politics was the contribution of these activists to creative city politics—similar to difficulties Spanish activists are facing today due to their alliances with local authorities in various municipalities, and their contribution to the city branding (Darly and McClintock 2017; Simon-Rojo et al. 2018).

In Athens, this tendency was especially remarkable after the arrival of thousands of migrants to Athens, who together with activists set up various spaces for housing, food provision and health care, among other needs. After these mobilisations, the municipality of Athens took on the brand of Solidarity City (see Solidarity Cities), hosting various events, from conferences to art festivals, and entering into global competition with other cities.

Conclusions: Urban Food Activism in Athens in the Everyday Life of Food Scarcities

Urban activism in Athens between the mobilisations of 2008 and 2015, and urban food activism more precisely, was reshaped particularly in ways that can be designated as non-state-centric, autonomous, everyday life or more invisible politics and repertories. Features or categories which are more common in the urban peripheries (Lefebvre 2009; Gutierrez 2017). As mentioned in the previous sections, and agreeing with Giorgio Agamben (2015) and Federici and Sitrin (2016), the more 'repressive' role of the state and the 'absolute liberal paradigm in the economy', particularly contributed to reshaping and disseminating these activist politics and repertories more widely.

Their efforts towards the de-commodification and the direct and collective organisation of social reproduction, the caregiving and the food provisioning, frequently contributed to embedding their activities in those needed for the everyday reproduction of lives. Activities, such as the caregiving and the food provisioning, whose lack of recognition and visibility feminist activists and scholars have been denouncing (Federici 2012; see Viewpoint 2015; Bhattacharya 2017). With no demands of local authorities, and through militant–activist work, activist repertories acquired a more direct, inclusive and permanent dimension: by frequently squatting empty public buildings, abandoned, thanks to the dismantling of the welfare system; by performing prefigurative politics which replaced the more hierarchical forms of organisation with assemblies; by building direct solidarity relationships between consumers and farmers, with no intermediaries (no NGOs, no political parties, no food industry); and also by building anti-fascist relationships and activist repertories beyond the more confrontational ones and in everyday life through non-racist, non-sexist, horizontal open spaces.

Essentially, the important relevance these forms of activism acquired in meeting everyday needs, when care and social reproduction are in times of crisis, certainly contributed with new repertoires to the 'politics in feminine' (Gutierrez 2017, p. 69). They questioned as well the categories generally assigned these forms of activism as being 'weak', 'spontaneous' or 'informal' (Leontidou 2010; Petropoulou 2014). But, agreeing with Judith Butler, the category we certainly can assign these forms of activism is their precarity. The pressures they regularly faced due to their non-state-centric politics and repertories were either repressed or unrecognised (Dalakoglou 2012, p. 539; Markantonatou 2015) or were pushed to replace their more autonomous and antagonistic forms of organisation into those that are more institutional, such as the social and solidarity economies recognised today by activists and local authorities and supported and promoted by specific legislation. Agreeing with Margit Mayer (2013), these pressures create more cooperative relationships between activists and local authorities by offering innovative solutions to solve problems, such as food scarcity, opened various tensions. As we observed in Athens, it created divisions. Furthermore, as we observed after 2015 especially, they contributed to the city branding strategy and the neoliberalisation of the city.

With their antagonistic features, they distanced themselves from other progressive food activists—those within the categories of food justice and food sovereignty that frequently adopt a more cooperative and demanding relationship with their local states (Holt-Gimenez and Shattuk 2011). As we observed today in various cities from these same southern European peripheries, by demanding the 'right to food' or the 'human right to food', they try to open spaces of negotiation with local authorities and build 'sustainable healthy inclusive and democratic food systems'.

References

Agamben, G. (2015). *From the State of Control to a Praxis of Destituent Power.* https://roarmag.org/essays/agamben-destituent-power-democracy/. Accessed 26 January 2018.

Arampatzi, A. (2012). The Urban Roots of Anti-Neoliberal Social Movements: The Case of Athens, Greece. *Environment and Planning A, 44*, 2591–10, https://doi.org/10.1068/a44416.

Arampatzi, A. (2017). The Spatiality of Counter-Austerity Politics in Athens, Greece: Emergent "Urban Solidarity Spaces" [Special Issue Interrogating Urban Crisis: Governance, Contestation and Critique]. *Urban Studies, 54*(9), 2155–2171.

Armstrong, A. (2011, August–September). Insolvent Futures/Bonds of Struggle [Special Issue on Debt]. *Reclamations Journal, 4*, 4–7.

Armstrong, A. (2015, August). Infrastructures of Injury. *Lies Journal, 2*, 130–135.

Baker, S., Kousis, M., Richardson, D., & Young, S. (Eds.). (1997). *The Politics of Sustainable Development. Theory, Policy and Practice Within the European Union.* London and New York: Routledge.

Bhattacharya, T. (Ed.). (2017). *Social Reproduction Theory Remapping Class, Recentering Oppression*. London: Pluto Press.

Butler, J. (2015). Notes Toward a Performative Theory of Assembly. Cambridge: Harvard University Press. ISBN 9780674967755.

Carabancheleando. (2017). *Diccionario de las Periferias. Métodos y saberes autónomos desde los barrios*. Madrid: Traficantes de sueños.

Castells, M. (1983). *The City and the Grassroots. A Cross-Cultural Theory of Urban Social Movements*. Berkeley: University of California Press.

Castoriadis, C. (1991). *Philosophy, Politics, Autonomy*. Oxford: Oxford University Press.

Colomb, C. (2012). Pushing the Urban Frontier: Temporary Uses of Space, City Marketing, and the Creative City Discourse in 2000s Berlin. *Journal of Urban Affairs, 34*(2), 131–152.

Dalakoglou, D. (2012). Beyond Spontaneity. *City: Analysis of Urban Trends, Culture, Theory, Policy, Action, 16*(5), 535–545. https://doi.org/10.1080/13604813.2012.720760.

Dalakoglou, D. (2013a). The Crisis Before the Crisis: Violence and Urban Neoliberalisation in Athens. *Social Justice, 39*(1).

Dalakoglou, D. (2013b). From the Bottom of the Aegean Sea'1 to Golden Dawn: Security, Xenophobia, and the Politics of Hate in Greece. *Studies in Ethnicity and Nationalism, 13*(3).

Darly, S., & McClintock, N. (2017). Introduction to Urban Agriculture in the Neoliberal City. *ACME: An International Journal for Critical Geographies, 16*(2), 224–231.

della Porta, D. (2017). *Rethinking Greece: Donatella Della Porta on Social Movements and Electoral Democracy in Times of Austerity*. http://www.greeknewsagenda.gr/index.php/interviews/rethinking-greece/6385-rethinking-greece-donatella-della-porta. Accessed 31 December 2017.

della Porta, D., Andretta, M., Fernandes, T., O'Connor, F., Romanos, E., & Vogiatzoglu, M. (2016). *Late Neoliberalism and Its Discontents in the Economic Crisis: Comparing Social Movements in the European Periphery*. Basingstoke: Palgrave Macmillan. ISBN 97833195350790.

Federici, S. (2004). *Caliban and the Witch: Women, the Body, and Primitive Accumulation*. New York: Autonomedia.

Federici, S. (2012). *Revolution at Point Zero. Housework, Reproduction and Feminist Struggle*. Oakland: PM Press. Creative Commons.

Federici, S., & Sitrin, M. (2016). Social Reproduction: Between the Wage and the Commons. *Roar Magazine, 2*. https://roarmag.org/magazine/social-reproduction-between-the-wage-and-the-commons/. Accessed 31 December 2017.

Fraser, N. (2017). Crisis of Care? On the Social-Reproductive Contradictions of Contemporary Capitalism. In T. Bhattacharya (Ed.), *Social Reproduction Theory Remapping Class, Recentering Oppression*. London: Pluto Press.

Graeber, D. (2002). The New Anarchists. *New Left Review, 13*, 61–73.
Gutierrez Aguilar, R. (2015). Mujeres, reproducción social y luchas por lo común. Ecos de la visita de Silvia Federici a México en otoño del 2013. *Bajo el Volcán, 15*(22), *marzo-agosto* (pp. 63–69). México: Benemérita Universidad Autónoma de Puebla Puebla.
Gutierrez Aguilar, R. (2017). *Horizontes comunitario-populares. Produccion de lo comun mas alla de las politicas estado-centricas.* Madrid: Traficantes de suenos.
Hadjimichalis, C. (2014). Crisis and Land Dispossession in Greece as Part of the Global Land Fever. *City: Analysis of Urban Trends, Culture, Theory, Policy, Action, 18*(4–5), 502–508.
Hadjimichalis, C., & Hudson, R. (2014). Contemporary Crisis Across Europe and the Crisis of Regional Development Theories. *Regional Studies, 48*(1), 208–218.
Holloway, J. (2005). Zapatismo as Political and Cultural Practice. *Humboldt Journal of Social Relations, 29*(1), 168–178.
Holloway, J. (2010). *Crack Capitalism.* London: Pluto Press.
Holt Giménez, E., & Shattuck, A. (2011). Food Crises, Food Regimes and Food Movements: Rumblings of Reform or Tides of Transformation? *The Journal of Peasant Studies, 38*(1), 109–144. https://doi.org/10.1080/03066150.2010.538578.
Katsarova, R. (2015). *Repression and Resistance on the Terrain of Social Reproduction: Historical Trajectories, Contemporary Openings.* https://www.viewpointmag.com/2015/10/31/repression-and-resistance-on-the-terrain-of-social-reproduction-historical-trajectories-contemporary-openings/. Accessed 31 December 2017.
Koronaiou, A., & Sakellariou, A. (2013). Reflections on "Golden Dawn", Community Organizing and Nationalist Solidarity: Helping (only) Greeks. *Community Development Journal, 48*(2), 332–338.
Kritidis, G. (2014). The Rise and Crisis of the Anarchist and Libertarian Movement in Greece, 1973–2012. In B. van der Steen, A. Katzeff, & L. van Hoogenhuijze (Eds.), *The City Is Ours: Squatting and Autonomous Movements in Europe from the 1970s to the Present.* Oakland: PM Press.
Lefebvre, H. (2009). State, Space, World. In N. Brenner & S. Elden (Eds.), *Selected Essays.* Minneapolis: Regents of the University of Minnesota.
Lekakis, J., & Kousis, M. (2013). Economic Crisis, Troika and the Environment in Greece. *South European Society and Politics, 18*(3), 305–331. https://doi.org/10.1080/13608746.2013.799731.
Leontidou, L. (1990). *The Mediterranean City in Transition: Social Change and Urban Development.* Cambridge: Cambridge University Press.
Leontidou, L. (2010). Urban Social Movements in "Weak" Civil Societies: The Right to the City and Cosmopolitan Activism in Southern Europe. *Urban Studies, 47*(6), 1179–1203.

Leontidou, L. (2014). The Crisis and Its Discourses: Quasi-Orientalist Attacks on Mediterranean Urban Spontaneity, Informality and Joie de Vivre. *City, 18*(4–5), 551–562.

Makrygianni, V., & Tsavdaroglou, H. (2011). Urban Planning and Revolt: A Spatial Analysis of the December 2008 Uprising in Athens. In A. Vradis & D. Dalakoglou (Eds.), *Revolt and Crisis in Greece. Between a Present Yet to Pass and a Future Still to Come*. Oakland, Baltimore, Edinburgh, London and Athens: AK Press and Occupied London.

Markantonatou, M. (2015). State Repression, Social Resistance and the Politicization of Public Space in Greece Under Fiscal Adjustment. In F. Eckardt & J. R. Sanchez (Eds.), *City of Crisis. The Multiple Contestation in Southern European Cities Urban Studies*. Bielefeld: Transcript Verlag.

Markusen, A. (2006). Urban Development and the Politics of a Creative Class. *Environment & Planning A, 38*, 1921–1940.

Mayer, M. (2006). Manuel Castells' the City and the Grassroots. *International Journal of Urban and Regional Research, 30*(1), 202–206.

Mayer, M. (2013). First World Urban Activism. Beyond Austerity Urbanism and Creative City Politics. *City: Analysis of Urban Trends, Culture, Theory, Policy, Action, 17*(1), 5–19.

Mayer, M. (2016). Neoliberal Urbanism and Uprisings Across Europe. In M. Mayer & H. Thorn (Eds.), *Urban Uprisings: Challenging Neoliberal Urbanism in Europe*. Basingstoke: Palgrave Macmillan.

Mayer, M., Thorn, C., & Thorn, H. (Eds.). (2016). *Urban Uprisings: Challenging Neoliberal Urbanism in Europe*. Basingstoke: Palgrave Macmillan.

McMichael, P. (2009). A Food Regime Genealogy. *The Journal of Peasant Studies, 36*(1), 139–169.

Milan Urban Policy Pact. https://www.milanurbanfoodpolicypact.org/. Accessed January 2018.

Mohandesi, S., & Teitelman, E. (2017). Without Reserves. In T. Bhattacharya (Ed.), *Social Reproduction Theory Remapping Class, Recentering Oppression*. London: Pluto Press.

Nadasen, P. (2012). *Rethinking the Welfare Rights Movement*. New York: Routledge.

Nasioka, K. (2014). Communities of Crisis: Ruptures as Common Ties During Class Struggles in Greece, 2011–2012. *South Atlantic Quarterly (2014), 113*(2), 285–297. https://doi.org/10.1215/00382876-2643621.

Nasioka, K. (2017). *Ciudades en Insurrección. Oaxaca 2006/Atenas 2008*. Guadalajara: Cátedra Interinstitucional Universidad de Guadalajara-CIESAS-Jorge Alonso.

Newman, S. (2007). Anarchism, Poststructuralism and the Future of Radical Politics. *Substance, 36*(2), 3–19. ISSN 0049-2426. https://doi.org/10.1353/sub.2007.0035.

Newman, S. (2011). Postanarchism and Space: Revolutionary Fantasies and Autonomous Zones. *Planning Theory, 10*(4), 344–365. https://doi.org/10.1177/1473095211413753.

Newman, S. (2016). What is an Insurrection? Destituent Power and Ontological Anarchy in Agamben and Stirner. *Political Studies*. https://doi.org/10.1177/0032321716654498.

O'Brien, T. (2017). Urban Movements in Neoliberal Europe. *Social Movement Studies*. https://doi.org/10.1080/14742837.2017.1393410.

Pastor, J. (2011). El Movimiento 15-M. Un nuevo actor sociopolítico frente a la dictadura de los mercados. *Sociedad y Utopía, 38,* 109–124.

Petropoulou, C. (2014). Crisis, Right to the City Movements and the Question of Spontaneity: Athens and Mexico City. *City, 18*(4–5), 563–572.

Rakopoulos, T. (2014). The Crisis Seen from Below, Within, and Against: From Solidarity Economy to Food Distribution Cooperatives in Greece. *Dialect Anthropology, 38,* 189–207.

Scott, J. C. (1985). *Weapons of the Weak. Everyday Forms of Peasant Resistance* (p. 389). New Haven and London: Yale University Press.

Shaw, L. H. (1969). *Postwar Growth in Greek Agricultural Production*. Special Studies Series 2. Athens, Greece: Center of Planning and Economic Research.

Simon-Rojo, M., Morales-Bernardos, I., & Sanz-Landaluze, J. (2018). Food Movements Oscillating Between Autonomy and the Co-Production of Public Policies. Lessons from Madrid. *Nature and Culture, 13*(1). ISSN 1558-6073 [Print] and ISSN: 1558-5468.

Solidarity Cities. http://solidaritycities.eu/. Accessed 31 December 2017.

Springer, S., Ince, A., Pickerill, J., Brown, G., & Barker, A. J. (2012). Anarchist Geographies. *Antipode: A Radical Journal of Geography, 44*(5): 1579L1754.

Stavrides, S. (2014). Emerging Common Spaces as a Challenge to the City of Crisis. *City, 18*(4–5), 546–550.

Urbact. http://urbact.eu/. Accessed 31 December 2017.

Vasilopoulou, S., & Halikiopoulou, D. (2015). *The Golden Dawn's 'Nationalist Solution': Explaining the Rise of the Far Right in Greece*. New York: Palgrave Macmillan.

Viewpoint Magazine. (2015). *Social Reproduction, Issue 5*. https://www.viewpointmag.com/2015/11/02/issue-5-social-reproduction/. Accessed 31 December 2017.

Vradis, A., & Dalakoglou, D. (Eds.). (2011). *Revolt and Crisis in Greece. Between a Present Yet to Pass and a Future Still to Come*. Oakland, Baltimore, Edinburgh, London and Athens: AK Press and Occupied London.

VVAAs. (2011). *Pensar las Autonomias. Alternativas de emancipacion al capital y el Estado*. México: Bajo Tierra Ediciones-Sisifo Ediciones.

Zechner, M., & Rübner Hansen, B. (2015). *Social Reproduction and Collective Care*. https://theoccupiedtimes.org/?p=14000. Accessed January 2018.

CHAPTER 5

From Obedience to Resistance: Understanding Equal Rights to Education Movement as a Cultural Process

Anqi Liu

Introduction

The current study is an ethnographic one adopting a cultural perspective in analysing the Equal Rights to Education Movement, a self-organised grass-roots movement aiming at the elimination of *hukou* restriction in education access in China. A grounded theory approach was chosen to unpack the transformation process from obedience to resistance embodied in the movement participants. Through empirical data gathered by in-depth interviews with movement participants as well as observations and analysis of movement-generated content, policies and media coverage, the study underlines the complex internal dynamics and impacts involved in this particular movement. It is during this unpacking that the role of social movement as a cultural process involving construction of movement subculture and production of meaning and knowledge as

A. Liu (✉)
Albert-Ludwigs-Universität Freiburg, Freiburg im Breisgau, Germany

N. M. Yip et al. (eds.),
Contested Cities and Urban Activism, The Contemporary City,
https://doi.org/10.1007/978-981-13-1730-9_5

well as practising of dealing with opponents and internal disputes is thus highlighted.

On Equal Rights to Education Movement

Equal Rights to Education Movement is a loosely associated nation-wide grass-roots movement, which aims to eliminate household registration (*hukou*) restriction on college entrance and to end the unjust distribution of education resources. It originated in Beijing in 2010 and rapidly gained momentum, having corresponding movements in major host cities for domestic migrants such as Shanghai and Guangzhou. The movement was initiated by Xu Zhiyong, a prominent legal scholar and civil rights lawyer and founder of 'Rights-defending (*weiquan*) Movement' and New Citizen Movement; the latter hopes to facilitate a peaceful transition of the country towards constitutionalism, from a 'servants' society' (*Chenmin Shehui*) towards a civil society (*Gongmin Shehui*) ('Xu Zhiyong and New Citizen Movement' 2013).

The movement took place in a context of massive domestic migrants. According to the Population Census in 2010, currently there are 260 million out of the whole population in China whose residence is different from their household registration. Excluding those that still reside in the same city as they registered (only in different districts), the floating population still numbers as high as 220 million (National Bureau of Statistics of China 2010).

The participants of the movement are 'domestic migrants' in no possession of host city's *hukou*, whose children are therefore not entitled to local education in most cases. The age of the participants' children varies from primary school entry level (five or six years old) to *gaokao* age (around 18 years old). They are from different walks of life, among whom there is a considerable proportion of the loosely defined emerging middle class in China, especially among the active members.

The campaign started with four parent volunteers in Beijing in 2010 and grew to more than 100,000 participants in 2012. With the sentence of the main founder of the movement Xu Zhiyong in the beginning of 2014, the movement has been in a less salient phase. By the time when the research was conducted in summer 2014, few activities were taking place except for one court case against Beijing Education Examinations Authority.

Fig. 5.1 Activists holding books of gathered signatures in front of Ministry of Education, 26 April 2012 (*Source* Picture taken by Ding Senxing with permission to reproduce)

Participants organised petitions to government bodies (Ministry of Education and its local branches), gathered signatures on the street, held seminars with right-defending lawyers and scholars, met representatives of the National People's Congress to raise their cause and actively increased the publicity through both mainstream media and social media. Participants were organised through QQ and wechat (Chinese messaging apps) groups, online forums, and formal and informal gatherings offline (Fig. 5.1).

Certain policy changes have been achieved, one of them being in 2010 which granted those migrant children with 'five documents' equal rights to primary and middle school education as those with Beijing *hukou*, which is considered a response to the gathering of more than 2200 signatures as well as a parents' demonstration in front of the Educational Committee of Haidian District. However, most of the campaign activities were met with unwelcoming gestures from the state. In July 2013, main founder of the movement Xu Zhiyong was officially detained following a three-month unofficial house arrest. This announced the beginning of slamming-down of the movement, with more frequent questioning, house raid and mobility limitation of major participants, closing of main websites and tighter surveillance on QQ groups and other social media accounts related to the movement. Shortly after, Xu was sentenced to four years for 'amassing a crowd to disturb order in a public space'.

Besides the movement participants and the state, there was another group of actors involved in the movement: counter-activists. They were Beijing *hukou* holders that were actively engaged in opposing the claims of Equal Rights to Education Movement. Their main arguments were that 'outsiders' should go back to their hometown, rather than competing for resources in Beijing. There were heated debates online. In a few cases, the online contention took its offline form-interrupting petition in front of Ministry of Education, '*renrou*' (revealing personal information), stalking and threatening of activists. Unfriendly attitudes from the side of counter-activists were met with varied responses from the activists. Some of them swore back with great resentment, both online and offline, while some tried to engage in more rational dialogues.

Social Movement as a Cultural Process

Despite the persistent uncertainty of results (Rucht 1992), studies on the impacts of social movements abound. Beside recent studies on economic effects and biographical consequences, scholars have mainly focused on the political and cultural dimensions. However, as Giugni (1998) has pointed out: a striking disparity exists between the large body of work on political outcomes and the sporadic studies on the cultural impacts of social movements. Admittedly, some authors have worked on social movements' cultural effects: such as spillover effects from one movement to the other (Meyer and Whittier 1994), social movement's capacity to generate social capitals (Diani 1997) and their impact of change in social values (Rochon and Mazmanian 1993). Amenta and Young (1999) have also showed a range of cultural effects, including 'transformations in belief systems or ideologies, new collective identities, innovative action repertoires, impacts on material culture, and influences on the practices and culture of mainstream institutions'. Klandermans (Johnston and Klandermans 1995) distinguishes three different processes of meaning construction in the movement context: public discourse, persuasive communication and consciousness raising during episodes of collective mobilisation. Through articulation of the notion 'ideoculture', Fine (1995) has highlighted the role of groups as locale of cultural enactment, where values and emotions were socialised. These studies are still scare in comparison with the huge amount of works on political outcomes: largely what Gamson (1975) terms as acceptance and new advantages.

Morris (1992, p. 35) has argued that culture must be brought back into social movement analysis if such analysis is to be free of structural determinism. Diani (1997, p. 135) has defined cultural outcome as 'a broader set of actions meant to shape the processes by which contemporary societies produce and reproduce moral standards, information, knowledge, and life practices'. Klandermans (1989, p. 9) has highlighted the importance of problem framing by arguing that interpretations of the situations instead of reality 'guide political actions', thus the role of social movement organisations as 'sponsors of meaning' (Klandermans 1989, p. 10). Most research concerning movements and culture also sees movements as vehicles for cultural changes (Amenta and Young 1999), that struggles take place not just in confrontations with the state, but in the building of culture as well. It is imperative that more attention be paid to social movements' nature of culture-making in order to comprehend its multidimensional nature. As processes of cultural change may occur over long periods, making it difficult to be captured by social scientists (Putnam 1993, p. 318), thick descriptions of the movement's dynamics might be of service in such studies.

McAdam et al. (2001) have called for a conceptual shift, away from looking for invariant causes and effects to looking for mechanisms and processes. della Porta (2014) also notes that there is no way to trace outcomes of complex social processes such as social movements without having robust descriptions and explanations of their operations. Therefore, a constructivist approach is adopted in this study, where dichotomies such as rational versus emotional, political versus cultural were bridged. An effort of reconstruction and translation is made to understand the meanings actors attach to the movement. It is argued that not only the cultural has an impact on the political, they are also intricately intertwined concepts. The changing political culture embodied in the movement participants is an anchor of the current study. The movement is treated culturally, and this approach allows for the investigation of the processes by which social movements are interpreted by members.

Contexualisation and Relevance of the Current Study

Unwilling to 'recognize citizens' political rights to enable individuals to protect themselves against the state' (Goldman 2002, p. 159), the Chinese state 'patrols the boundary of social activism with periodic, selective and at times violent repressions' (Hsing and Lee 2010). In fact, since the end of the Great Proletarian Cultural Revolution, social movements that tried to mobilise the Chinese masses have met a similar fate: suppression and de-legitimisation. From the perspective of individual citizens, 'uncertainty and self censorship' (O'Brien and Stern 2008, p. 24) have played a key role in discouraging Chinese contentions. With the considerable power and authority over society wielded from the state, social movements in China are 'forced to keep a low profile or to lead a semi-clandestine existence' (Ho 2008, p. 3). Current studies on social movements and activism in China have mostly been attempting to unpack political opportunity structures and/or resource mobilisations (Ho 2008; Hsing and Lee 2010; Perry 1985; Zhao 2001). Some other scholars have concerned themselves with contentious repertoire employed in movements (Perry 2002; Yang 2008, 2009; Tarrow 2008; O'Brien and Li 2006; Jin 2011) Few studies have engaged in the investigation of the impacts of social movements: with Xia and Guan (2014) and O'Brien and Li (2006) as rare exceptions. There is especially a scarcity of studies based on case studies or those with an in-depth, 'from-below' approach, or those with an innovative focus on dimensions such as emotions and cultural impacts other than opportunity structure and resource mobilisation. Having gained unique access into the field, the current study attempts to adapt an ethnographic approach, focusing on participants‘ lived experiences and transformed meanings and consciousness, giving voice to notable silence in literature, documenting and analysing the Equal Rights to Education Movement as a cultural process.

The very spread and depth of social movement studies have also contributed to fragmentation in the field. Focusing on the West and the North is still largely prevalent in most social movement theories (della Porta 2014). In non-democratic countries characterised by degrees of coercion and rare openings for mobilisation, innovative perspectives are required to deepen the understanding of local social movements. At the same time, the collective mobilisation and insurgent waves from Eastern Europe, Asia to North Africa have been addressed mainly by area studies, focusing very much on each area's particularity. This gap between different disciplines urged author of this article to bring together different paradigms and to search for insights that 'both "live" in particular settings' (Edwards 2011, p. 6) and have something to contribute to social movement studies across contexts. By placing Chinese social movement in a comparative framework, the study is an attempt to find reciprocal contacts among the various streams of literature on social movement and Chinese studies, and to seek balance between a circumspect, context-specific approach without going into exoticising China.

Research Methods

This research is an exploratory qualitative study that involved ethnographic approaches. It was based on the grounded theory method (Glaser and Strauss 1967), especially adopting its approach of simultaneous data collection and analysis as well as the constant comparison and reflection from where theories evolve. Admittedly, as della Porta and Rucht (2013, p. 12) have pointed out, ethnographic methods are not commonly used in social movement studies, which tend to prefer macro-characteristics such as political opportunities and resource mobilisation. To understand movements' mechanisms and processes, robust descriptions would be of service (McAdam et al. 2001; della Porta 2014; Putnam 1993). This method was adopted in the current study to reveal the micro-aspects of participants' lived experiences and the 'web of meaning' (Bray 2008, p. 301).

Primary data are the main source which is collected by the researcher from August 2014 to October 2014 in Beijing. Data were fretted out through multiple methods: interviews, non-participatory observations and reviews of movement-generated materials, participants' social media content as well as government policies. The access to the field was through Transition Institute, a civil society think tank that was closely related to New Citizen Movement and Rights-Defending Movement. Researchers from the Institute have been following the movement and also informally facilitated legal support for the movement; therefore, initial access and trust were granted. As the major source of data, non-participatory observations and in-depth interviews were conducted simultaneously, offering access and focusing points to each other. 'Theoretical sampling' (Glaser and Strauss 1967) was favoured to choose 'crucial' cases for in-depth interviews. Twelve in-depth interviews with key participants of the movement and two with family members of the activists were conducted. The interview guidelines comprised of three sets of questions concerning background of the family, children's education, and their involvement in the movement. The interviewees were to talk about what was important for them instead of being directed. This opened the researcher to unexpected topics such as emotional expressions of injury, violence in protest sites.

Non-participatory observations were performed to complement and testify findings from the interviews and to offer 'a humanist approach into the complexity of the culture and political world of people' (Bray 2008, p. 301). The researcher attempted to attend the two court case hearings in September but was stopped by the court police along with other activists that gathered outside the court but were not litigants of the lawsuit. This, however, turned out to be an opportunity to observe activists' interactions with court police and plain-cloth police. The researcher also participated in several informal gatherings such as dinner parties. In these occasions, the researcher was able to be exposed to unexpected information and interact with a bigger group of activists.

Aside from interviews and observations, movement-generated materials such as meeting minutes, public appeals, and website content were examined to gather information from the perspective of the movement. Through following the social media accounts of the participants, their views, attitudes, and political opinions were traced. The examination of media coverage and government policies located a context within which the movement took place. At the same time, expert interviews with scholars specialising in the field gave an overview and multiple understanding of this movement.

The process of data analysis was carried out at the same time during the data collection. Some sensitising concepts emerged in the process of data analysis, which became important clue for data collection. The analysis of interview transcripts and field notes was based on an inductive approach geared to identifying patterns in the data by means of thematic codes, where 'the patterns, themes, and categories of analysis emerge out of the data' (Patton 1980, p. 306). The coding followed Kozinets' (2010) procedure. In the first round of microanalysis, open codes such as 'anger and frustration' and 'feelings of exclusion' were given to the material; the axial coding process saw the associating of interrelations between basic concepts and the grouping of similar codes into categories such as 'emotions', 'notion of citizenship' and 'attitude change towards the state'. Some provisional hypothesis also arose during the line-by-line analysis: it was clear that participants of the movement experienced a certain change in their motivation, outlook, and practices throughout the process of their involvement in the movement. With the aid of theoretical input and systematic comparison were unpacked the relations between different categories and central themes including enlargement of moral motivation, awakening of rights consciousness, increasing autonomy, and process of practising deliberation, on which the following parts will elaborate.

Behind the Movement: Expansion of Moral Motivations

Most of the interviewees shared a common trajectory: in the beginning, they were only fighting for their own children's rights. It was a question of 'I worked and paid tax in Beijing, why can't my children go to school here?'; through involvement in the movement, they started paying attention to the sociopolitical context of their sufferings.

Private Experiences of Injury

'I am forced to fight, what would my child think of me if I haven't?' This is a quote from Lingyun when she explained why she joined the movement. She was not alone: interviewees perceived the education policy in Beijing to be 'inhumane', 'suffocating', 'a slap on the face', depriving them of basic rights and 'giving them no options'. Lingyun said in one of the September court hearings that 'the government says the children are the future of our country but this policy is nipping them in buds; it is an evil policy, and people carrying out this policy are holding the candle to the devil!' Hurt feelings were obvious when parent-activists recalled the family's experiences in such difficult times. Parents complained about 'restrictive accessibility'(Zaixin) and lack of 'even a little bit of justice' (Liutang). Children of all the interviewees experienced a phase of lack of motivation in studying. A few took actions such as tearing apart honorary certificates from school, skipping classes, and giving up on exams. The parents were apologetic for not being able to protect their children against injustice, or to grant them basic education.

Auyero (2004, p. 439) has dissected that 'protester's actions, thoughts, and feelings during the uprising were deeply informed by the history of his or her life'. Family histories of perceived discrimination and exclusion, and undeserved experience of injury had an important role in shaping the initial involvements in the movement. Taking the related educational policy based on *hukou* as responsible, one clear motivation behind mobilisation of Equal Rights to Education participants is what Honneth (1995, p. 163) calls 'moral experiences stemming from the violation of deeply rooted expectations regarding recognition'.

Forging of Collective Identity

As Honneth (1995) has pointed out, hurt feelings can become the motivational basis for collective resistance 'only if subjects are able to articulate them within an inter-subjective framework of interpretation that they can show to be typical for an entire group' (p. 162). Moral feelings of indignation have led some individuals to become more informed about the issue and the bigger political context. They discovered that their individual experiences of disrespect were shared by the vast numbers of residents without local *hukou* in Beijing. From this realisation arose a collective identity and an enlargement of motivation, from stopping personal pain to stopping injustice inflicted on the collective.

From the private experience of injury, the interviewees developed a sense of empathy towards others with similar problems and thus constructed affinities. 'There are 8,000,000 people without local *hukou* in Beijing and we all face this problem' said Liutang (33:24). Through the movement, the participants forged and demonstrated a 'we' that were in search of visibility. The struggle was not limited to personal concern but was brought to the level of fighting for recognition of the group and for others not to experience the same pain. This collective identity is best illustrated in the 12th Open Letter to Minister of Education from non-Beijing-*hukou* Parents: 'we are Chinese citizens living, working and paying tax in Beijing, without Beijing *hukou*. We are from all walks of life: construction, domestic service, entrepreneur, journalism, and lawyer. We act on behalf of the 30,000 others who signed the petition to fight for the educational rights of children of 8,000,000 domestic migrants in Beijing and two million domestic migrants all over the country'.

The collective identity has also produced solidarity and sustained motivation of participation, which got strengthened through the movement. By the time the interviews were conducted, the children of most interviewees were either studying outside of Beijing or abroad. Activists' children couldn't enjoy the benefit of any policy change even if it had happened. Yet, these parents were still actively taking part in the movement out of 'a sense of responsibility: parental responsibility, as well as social responsibility' (Lifeng), in hope that the painful experiences of their own family don't repeat in other families (Lifeng; Shujiang). They ask for 'a systematic solution that makes it possible for children to continue their studies near their parents' (Shujiang) and 'no more left-behind kids[1]' (Zaixin). This interpersonal solidarity has also facilitated the movement by offsetting the constraints of outside influences and countering shame and bolstering pride (Shemtov 2003). Facing suppression, activists exchanged contact information of their families and were ready for the worst to happen. All of the activists that were interviewed said that they would not give up despite all the difficulties and frustrations. Ruhai said 'there is this solidarity and tacit understanding among us, which makes me less fearful. After all, what's the big deal even if I am arrested'.

Through the Movement: From Subject to Citizen

Rights-Bearing Citizens: An Expanding Rights Consciousness

Although Chinese political thought has generally conceived of rights as concessions rather than as entitlements, that view has shown signs of change embodied in the interviewees. Participants of the current movement viewed that the government has a duty to realise citizens' rights, especially basic social rights such as education. They joined the movement objecting the violation of education rights. Through the movement, the idea of citizen rights has concreted and expanded.

T. H. Marshall (1964) most famously explored the notion of citizenship as status. In other words, citizenship endows the citizen with a status that involves access to various rights and powers. He points out that citizenship is comprised of three core elements: social, civil and political (Marshall 1964, p. 71). As above discussed, participants realised the violation of their education rights as basic social rights. They also felt deprived of economic welfare and social security included in Marshal's delineation of social citizenship. Many interviewees mentioned education as a due right of taxpayers. Moreover, most of the interviewees mentioned individual rights to property, personal liberty and justice,

[1] Children that are left behind in rural areas under the care of relatives when their parents emigrate to cities in seek of career opportunities.

components of civil rights. They especially became aware of the state's unscrupulous violations of these rights during the movement as they experienced arbitrary detentions and questionings as well as reckless censorship of online communications. After being detained and humiliated following one peaceful petition in front of Ministry of Education, one activist Qiaoxin wrote on weibo that 'my son, try to emigrate somewhere else. Mum hopes that you can live somewhere where no one can take away your dignity arbitrarily like they did to your mum'. Liutang said that through engaging in the court case and communication with scholars and lawyers, she reflected and realised that her citizen rights have been undergoing transgressions all the time, particularly by the state apparatus.

When it comes to political rights, Goldman (2002, p. 159) notes that in China 'political rights are to enable citizens to contribute to the state rather than to enable individuals to protect themselves against the state'. Although the movement was also largely framed in a depoliticised manner, activists endeavoured to exercise the right to participate in policy-making and realising the political citizenship. As Goldman and Perry (2002, p. 19) have noted, 'a growing sense of entitlement as exhibited through actions, more than words, suggest that important elements of the Chinese populace are adopting a definition of citizenship that includes the exercise of political rights'. Participants of the movement actively engaged in gathering signatures, contacting congress deputies and other legal methods to push towards a more participatory policy-making procedure. They no longer think that officials will automatically help solve problems, but realised the necessity to 'stand up for oneself'. Moreover, the action of actively practising freedom of association and assembly as granted by the constitution, though more often than not disregarded by state apparatus, was a firm demand in itself to realise political rights (Fig. 5.2).

The struggle for rights has more than a merely rhetorical impact on the participants' civic consciousness. As Foweraker and Landman (1997) noted, claims framed by movements in the frame of rights, citizenship and their political actions play a crucial function in cultivating citizenry.

Autonomous Citizens: De-Colonialisation of Life World

Habermas (1992) coined the term 'colonisation of the life world through power' to describe the internalisation of values indoctrinated by the state and co-opted media. Chinese citizenry is especially vulnerable

Fig. 5.2 Activist holding a sign that says 'We are all citizens of People's Republic of China', in front of Ministry of Education, 26 April 2012 (*Source* Picture taken by Ding Senxing with permission to reproduce)

to the proliferation of party-controlled ideologies. This colonisation of life and values, however, was to a certain extent offset through the movement. Not only do the activists consider themselves rights-bearing citizens, they also become increasingly autonomous, demarcating their distance from the state. Similar to what Goldman and Perry (2002, p. 12) have noticed in the transformation of the question what a responsible citizen entails in China since the 1980s, interviewees mentioned citizenship less in the collective discourse of seeking to strengthen the nation than emphasising protecting and enlarging their own individual and collective rights.

With little or no history of previous activism, majority of the activists had initial beliefs about a just and legitimate government. The movement brought forth 'a loss of trust in government agents', and participants have become increasingly disenchanted with authorities. Lingyun said she used to be 'patriotic' and 'protective when people criticised the government'. It was during her activism that Lingyun realised that 'many things really don't work right in China'. Activists like her started re-evaluating the state performance in its capacity of providing public good. Involvement in the movement has engendered an increased concern with public issues. Participants started with realising the violation of their own rights and went on to questioning the legitimacy of the state at large. They are critical about the 'unaccountability' (Wengong), 'hypocrisy' (Lingyun) (Shujiang), 'injustice' and 'gearing towards vested interest' (Youran) of the state and co-opted media and scholars. Content analysis from interviewees' social media also showed a critical shift in their political opinions. Activists become followers of liberal public intellectuals offering counter expertise against state-dominated information on social issues. With a partial success in breaking through the hegemonic information control, these self-claimed 'obedient citizens' would repost articles about disrespect of human rights and lack of freedom of expression and press, lack of rule of law, etc. During the Hong Kong Protest for universal suffrage in October 2014, for example, all the interviewees posted online commentary to show support. In their discussion among each other, such topics as 'constitutionalism' and 'democracy', 'human rights crisis in China' was frequently touched upon.

The participants had frequent interactions with police officials: police violence and brutality during protests, 'invitation for tea',[2] limited mobility and threats. Contrary to the common perception of the police as guardians of public good, many of the participants realised the violent nature of state apparatus. 'The police are like machines, stooges, they are violent and unscrupulous. They would beat up parents in front of their children. It's awful.' said Lingyun. A shift from feelings of trust, respect and intimidation towards public security authorities to feelings of distrust and contempt was an observable cultural process. Huarong said that she is now 'used to these intimidating tricks played by the police' and has no more fear. Fainted with shame when she was put under custody for the first time, Qiaoxin now said that 'those thrown into jail are usually people with good conscience, like Xu Zhiyong'. Yinglan described her interaction with the police as an absurd theatre play: 'they just simply repeat what they say, and most of the time they don't know what they are talking about'.

Another important shift in the participants' attitude is the doubt they started shedding on the monopolised official discourses. The accusation

[2] *Qing he cha* (invite to tea) euphemism for police interrogation. When invited to 'tea,' one is asked about their political activities and warned against further involvement.

of being 'stooges or tools of Western imperial forces' is a common one lobbed at social movement groups in non-Western developing countries (Flesher Fominaya 2010). Most of the interviewees were warned by the police not to be 'taken advantage of by people with special intentions'. However, activists took a distancing and ridiculing stance in countering these accusations. Lingyun told the police 'I would love to be used as long as my child can go to school'. Liutang countered it saying 'I wish to be used by some people, I am afraid I am not useful enough'. Yinglan said 'so-called citizenship awareness is within everybody. Once triggered, it will mature very fast. Human rights awareness doesn't need to be enlightened by anyone, what I do is inspired by the nature to protect my child. I don't need someone to take advantage of me'. Being unlawful and disturbing, social order was another accusation often inflicted upon activists. Lingyun used 'diaomin'[3] sarcastically to talk about herself. 'They would say that other parents can find ways around the policy, and that I should not fight against it either. Otherwise I am a 'diaomin' looking for troubles. All right, I am'. The movement produced transformed symbols, frames of meaning and norms that justify and dignify collective actions. The standing vis-à-vis the police changed over time and the official discourse discrediting the movement was met with suspicion, mocking if not complete antagonisation. Through the movement actors presented 'a distinctive subcultural profile' (Diani 1997, p. 137), the credibility of official media and the unshakability of the state legitimacy have altered and the all-encompassing matrix of party control over ideologies has faced challenges.

Within the Movement: Trial and Error in Practice

Edward Shils (1997, p. 337) has termed substantive civility as the 'readiness to moderate particular, individual or parochial interests and to give precedence to the common good'. In the case of the current movement, there are external and internal disputes that activists learn to deal with through trial and error. Therefore, the movement can be understood as a cognitive process of norm formation and activists facilitating themselves with the 'capacity to be able to act like a citizen' (Xia and Guan 2014, p. 418)

[3] Shrewd and unruly people, a tag often put on activists.

Externally: Deliberation and Compromising

Civic consciousness involves recognising each other as rights-bearing citizens, not only those that share the claims over similar rights, but also those having opposing interests: in the case of the current movement, the counter-activists. There was a huge amount of emotion-charged interactions both online and offline. In extreme cases, counter-activists waged threats to activists and their family's security. Despite the group's collective attempt to replace adversarial exchange with cooperative conversations and to engage in rational dialogues, the verbal violence online wasn't limited to only one direction. Activists of Equal Rights to Education Movement also have engaged in verbal insults towards counter-activists. However, the experienced participants all expressed understanding towards 'locals' that do not agree or support their struggles and tried to negotiate conflicts with less confrontational solutions. They said that they understand Beijing parents in fear of more fierce competition for their own children (Ruhai; Wengong). Huarong said that she tried to have a dialogue with counter-activists and that swearing back and forth is not helpful. She made several attempts to persuade counter-activists in leaving the protest site of activists and to make one-to-one dialogue with them afterwards.

Internally: Experimenting Self-Organisation and Internal Democracy

Equal Rights to Education Movement is a loosely connected voluntary association: the movement was held together by virtue of a network of personal connections and commitment to a common aim. The parents were to organise themselves based on the New Citizen Movement's principle of 'organising without organisation' (Caster 2014). The movement itself turned out to be a process of internal democracy building, where participants learned to organise themselves and mobilise consensus. Three major problems emerged: the existence of informal hierarchy, the disputes over whether partial inclusion is desirable and the divergence between hardliners who are for goal expansion and moderates who want to limit the movement to a depoliticised existence.

Internal democracy means no official hierarchy and a constant strive for symmetric communication and transparency on group matters (della Porta 2014). The decentralised nature of the movement with the tight censorship from the police made it difficult to insist on symmetric communication and transparency. Despite the absence of official hierarchy, there was a demarcation between core actors and peripheral participants. The level of political consciousness seemed to be the main dividing line between peripheral participants and core participants, which has led to disputes on the goal of the movement. Peripheral participants thus define goals in a more self-interested manner. There were complaints about certain activities for being too 'political'. Among the core participants, however, divergence was not so much about self-defined interests as much as about differences in personalities, approaches, etc. No formal positions of delegation and power were a challenge in a cultural context of passive or parochial political culture that relies heavily on top-down manners of decision-making.

To a large extent, participants were able to engage with each other across the lines of difference in order to fashion alliances and networks, as Huarong describes: 'whoever initiates something, the rest will support in whichever way possible, it's for the cause, not for any specific person'. The role of social movements as site of knowledge production where repertoires of mobilising consensus and deliberation and compromise evolved is therefore highlighted.

Conclusions

The current study is an ethnographic one based mainly on interviews and observations of key participants of the movement and review of movement-generated material. By documenting the ongoing processes through which the movement pursues the realisation of migrant children's rights to education and the recognition of migrants' citizen status in the city, an attempt is made to understand the internal dynamics of grass-roots activism. Through the analysis of empirical data in comparison with theoretical concepts, the internal dynamics of Equal Rights to Education Movement were unveiled; the transformed culture and meanings unpacked; and the role of social movement as a cultural process highlighted.

Throughout the movement, participants experienced a transformation in their outlook and an awakening of civic consciousness: from obedience to resistance, and from subject to citizen. It is elucidated that participants initially joined the movement out of emotional hurt feelings inflicted by a perceived unfair policy. The purpose was simply to fight for their children's right to local education. At the realisation that this demand was echoed with numerous others, a sense of solidarity and affinities developed based on a collective identity. The enlargement of motivation to the ending of injustice in general provided enduring incentives and offset external constraints long after the initial feelings of private injury. A clearer awareness of citizen rights and their violations and a more firm claim over the realisation of citizen rights were demonstrated. Participants had a transformed perception of the state and the state-individual relations: more critical view of the duties and legitimacy of the state, police and official discourse were observable among the interviewees. The movement itself also proved to be a site of learning to be 'civil': to compromise and to 'give precedence to the common good' (Shils 1997, p. 337). Over the course of the movement, internal democracy was constantly strived for and divergences among the participants were to a certain extent successfully bridged after some run-in period. Deliberation with external opponents was also a lesson learned through trial and error. As participants attached new meanings to concepts such as state, police and responsible citizenship, moral and ethical principles were redefined. By pluralising actors in culture formation, building social networks as well as providing opportunities to practise, the current movement became a possible and indeed legitimate site of culture forming and transforming. It is important to focus on the symbolic dimensions of collective mobilisation and its role as a process of culture-making. As Chu Zhaohui (14 September 2014) [Personal interview] has pointed out that only when more individuals change their aspirations of arriving at a higher social strata so as to enjoy the privileges of better education resources to aspirations of changing the unequal situation so that everyone can share education resources more equally, can real social changes possibly take place. The change under question is an ongoing cultural process among the participants of the movement, signified in the cultivation of citizenship consciousness. A subculture that honours collective action, solidarity, civic engagement and deliberation prevailed in the movement.

The discourse of Equal Rights to Education is far from anything revolutionary: activists didn't create their own rhetoric, but embraced values endorsed by certain part of the liberal elites and held a somewhat idealised picture of Western liberal democracies. The rhetoric of rightful resistance also revealed a hierarchical political imagination. Moreover, despite the attempt at experimenting horizontal organisation, existing internal democracy within the group is far from what della Porta and Rucht (2013, p. 225) described as symmetric, transparent, protective of minority rights, let alone 'critical of the state of contemporary democracies at large'. However, it is circumspect to contend that elements of a sprouting citizenship were observed when a hitherto obedient citizen culture was shaken. Being forced into an embedded existence might not bring immediately observable transformative changes, but can be instrumental to what Ho (2008) describes as 'incremental but certain changes'. And as Tarrow (2008) has pointed out, unintended responses to such incremental changes can be more effective than open challenges of the bases of political legitimacy. The cultural impacts of the movement embodied in the participants of the current study might be a drop in the ocean, but the preparation for a forthcoming solidarised, autonomous, engaged and 'civil' citizenry and a resistant political culture has to begin somewhere. As Xu Zhiyong (2014) wrote in his closing statement, 'The day will come when the 1.3 billion Chinese will stand up from their submissive state and grow to be proud and responsible citizens'.

References

Amenta, E., & Young, M. P. (1999). Democratic States and Social Movements: Theoretical Arguments and Hypotheses. *Social Problems, 57*, 153–168.

Auyero, J. (2004). When Everyday Life, Routine Politics, and Protest Meet. *Theory and Society, 33*(3–4), 417–441.

Beijing Municipal Commission of Education. (2014, April 18). *Suggestions of Educational Committee of Beijing on Enrolment of Students in Compulsory Education Period in 2014.* http://www.bjedu.gov.cn/publish/portal27/tab1654/info37238.htm.

Beijing Municipal Commission of Education, Beijing Municipal Comission of Development and Reform, Beijing Municipal Human Resources and Social Security Bureau and Beijing Municipal Public Security Bureau. (2012, December 29). *Plans for Children of Migrant Workers to Apply for Entrance Exams After Completion of Compulsory Education.* http://www.gov.cn/zwgk/2013-01/05/content_2305015.htm.

Bray, Z. (2008). Ethnographic Approaches. In D. Porta & M. Keating (Eds.), *Approaches and Methodologies in the Social Sciences: A Pluralist Perspective* (pp. 296–315). Cambridge and New York: Cambridge University Press.

Caster, M. (2014, June 6). *The Contentious Politics of China's New Citizens Movement.* https://www.opendemocracy.net/civilresistance/michael-caster/contentious-politics-of-china's-new-citizens-movement. Accessed 1 April 2015.

della Porta, D. (2014). *Mobilizing for Democracy: Comparing 1989 and 2011*. Oxford: Oxford University Press.

della Porta, D., & Rucht, D. (2013). Power and Democracy in Social Movements. In D. della Porta (Ed.), *Meeting Democracy: Power and Deliberation in Global Justice Movements*. Cambridge: Cambridge University Press.

Diani, M. (1997). Social Movements and Social Capital: A Network Perspective on Movement Outcomes. *Mobilization: An International Quarterly, 2*(2), 129–47.

Edwards. (2011). Introduction: Civil Society and the Geometry of Human Relations. In M. Edwards (Ed.), *The Oxford Handbook of Civil Society* (pp. 3–14). New York: Oxford University Press.

Fine, G. (1995). Public Narration and Group Culture: Discerning Discourse in Social Movements. In H. Johnston & B. Klandermans (Ed.), *Social Movements and Culture*. Minneapolis, MN: University of Minnesota Press.

Flesher Fominaya, C. (2010). Collective Identity in Social Movements: Central Concepts and Debates: Collective Identity in Social Movements. *Sociology Compass, 4*(6), 393–404.

Foweraker, J., & Landman, T. (1997). *Citizenship Rights and Social Movements: A Comparative and Statistical Analysis*. Oxford: Oxford University Press.

Gamson, W. (1975). *The Strategy of Social Protest*. Madison, IL: Dorsey Press.

Giugni, M. (1998). Was It Worth The Effort? The Outcomes and Consequences of Social Movements. *Annual Review of Sociology, 24*, 371–393.

Glaser, B., & Strauss, A. (1967). *The Discovery of Grounded Theory: Strategies for Qualitative Research*. Chicago: Aldine Publisher.

Goldman, M. (2002). The Reassertion of Political Citizenship in the Post-Mao Era: The Democracy Wall Movement. In M. Goldman & E. Perry (Ed.), *Changing Meanings of Citizenship in Modern China*. Cambridge, MA: Harvard University Press.

Habermas, J. (1992). Further Reflections on the Public Sphere. In C. Craig (Ed.), *Habermas and the Public Sphere* (pp. 421–461). Cambridge: The MIT Press.

Ho, P. (2008). Embedded Activism and Political Change in a Semiauthoritarian Context. In P. Ho (Ed.), *China's Embedded Activism: Opportunities and Constraints of a Social Movement*. London: Routledge.

Honneth, A. (1995). Disrespect and Resistance: The Moral Logic of Social Conflicts. *The Struggle for Recognition: The Moral Grammar of Social Conflicts* (pp. 160–170). Cambridge: Polity Press.

Hsing, Y., & Lee, C. (2010). *Reclaiming Chinese Society: The New Social Activism*. London: Routledge.

Jin, J. (2011). Institutionalized Official Hostility and Protest Leader Logic: A Long-Term Chinese Peasants Collective Protest at Dahe Dam in the 1980s. In J. Broadbent & V. Brockman (Eds.), *East Asian Social Movements Power, Protest, and Change in a Dynamic Region* (pp. 413–436). New York: Springer.

Johnston, H., & Klandermans, B. (1995). The Cultural Analysis of Social Movements. In H. Johnston & B. Klandermans (Ed.), *Social Movements and Culture* (Minneapolis, MN: University of Minnesota Press).

Klandermans, B. (1989). Introduction: Social Movement Organizations and the Study of Social Movements. In B. Klandermans (Ed.), *Organizing for Change: Social Movement Organizations in Europe and the United States* (Vol. 2). Greenwich, CT: JAI Press.

Kozinets, R. (2010). *Netnography: Doing Ethnographic Research Online*. Sage: Los Angeles, CA.

Marshall, T. H. (1964). *Class, Citizenship, and Social Development; Essays*. Garden City, NY: Doubleday.

McAdam, D., Tarrow, S., & Tilly, C. (2001). *Dynamics of Contention*. Cambridge: Cambridge University Press.

Meyer, D. S., & Whittier, N. (1994). Social Movement Spillover. *Social Problems, 41*(2), 277–298.

Morris, A. (1992). Political Consciousness and Collective Action. In A. Morris & C. Mueller (Ed.), *Frontiers in Social Movement Theory*. New Haven, CT: Yale University Press.

National Bureau of Statistics of China. (2010). *2010 Population Census*. http://www.stats.gov.cn.

New Citizens' Movement (China). (n.d.). http://self.gutenberg.org/articles/New_Citizens'_Movement_(China)#Equal_rights_for_education. Accessed 3 March 2015.

O'Brien, K., & Li, L. (2006). *Rightful Resistance in Rural China*. Cambridge: Cambridge University Press.

O'Brien, K. J., & Stern, R. E. (2008). Studying Contention in Contemporary China (July 7, 2007). In Kevin J. O'Brien (Ed.), *Popular Protest in China* (pp. 11–25, 219–225). Harvard University Press.

Patton, M. (1980). *Qualitative Evaluation Methods*. Thousand Oaks, CA: Sage Publications).

Perry, E. (1985). Rural Violence in Socialist China. *The China Quarterly, 103*, 414–440.

Perry, E. (2002). *Challenging the Mandate of Heaven: Social Protest and State Power in China*. Armonk, NY: M. E. Sharpe.

Putnam, R. (1993). *Making Democracy Work: Civic Traditions in Modern Italy*. Princeton, NJ: Princeton University Press.

Rochon, T., & Mazmanian, D. (1993). Social Movements and the Policy Process. *The Annals of the American Academy of Political and Social Science, 528*, 75–87.

Rucht, D. (1992). *Studying the Effects of Social Movements: Conceptualizations and Problems*. ECPR Joint Sessions, Limerick.

Shemtov, R. (2003). Social Networks and Sustained Activism in Local NIMBY Campaigns. *Sociological Forum, 19*(2), 215–244.
Shils, E. (1997). *The Virtue of Civility: Selected Essays on Liberalism, Tradition, and Civil Society.* Indianapolis: Liberty Fund.
Tarrow, S. (2008). The New Contentious Politics in China: Poor and Blank or Rich and Complex. In K. O'Brien (Ed.), *Popular Protest in China* (pp. 1–10). Cambridge, MA: Harvard University Press.
Xia, Y., & Guan, B. (2014). The Politics of Citizenship Formation: Homeowners' Collective Action in Urban Beijing. *Journal of Chinese Political Science, 19*(4), 405–419.
Xu, Z. (2014, January 23). *For Freedom, Justice and Love—My Closing Statement to the Court.* http://chinachange.org/2014/01/23/for-freedom-justice-and-love-my-closing-statement-to-the-court/. Accessed 20 April 2015.
'Xu Zhiyong and New Citizen Movement'. (2013, July 29). Radio Free Asia. http://www.rfa.org/mandarin/zhuanlan/zhongguotoushi/panel-07292013105036.html.
Yang, G. (2008). Contention in Cyberspace. In K. O'Brien (Ed.) *Popular Protest in China* (pp. 126–143) Cambridge, MA: Harvard University Press.
Yang, G. (2009). *The Power of the Internet in China: Citizen Activism Online.* New York: Columbia University Press.
Zhao, D. (2001). *The Power of Tiananmen State-Society Relations and the 1989 Beijing Student Movement.* Chicago: University of Chicago Press.

PART III

Urban Activism—Activists and Their Networks

CHAPTER 6

Has Urban Cycling Improved in Hong Kong? A Sociopolitical Analysis of Cycling Advocacy Activists' Contributions and Dilemmas

Hongze Tan and Miguel Angel Martínez López

Introduction

In recent years, bicycles, as a mode of urban transportation, have become increasingly visible in the political agenda and the public debate in Hong Kong, a traditionally 'non-cycling' city (Zhao 2010). Although the transportation issue as a whole, and the urban cycling issue in particular, is traditionally a 'public' area in which the government plays a key role, a significant change that recently occurred is that some non-governmental actors increasingly enjoyed active and influential roles in bringing changes to the issue. This article concerns one category of emerging

H. Tan (✉)
Nankai University, Tianjin, China

M. A. Martínez López
Uppsala University, Uppsala, Sweden
e-mail: miguel.martinez@ibf.uu.se

N. M. Yip et al. (eds.),
Contested Cities and Urban Activism, The Contemporary City,
https://doi.org/10.1007/978-981-13-1730-9_6

non-governmental actors in Hong Kong-cycling advocacy activists—to explore how they bring changes to both the cycling issue and the power relations surrounding it through interactions with other actors, especially state agencies. Besides, we will also reveal and discuss the real and potential dilemmas they face.

Hong Kong has been a Special Administrative Region of China since 1997 and was previously a British colony. It hosts a residential population of 7.23 million (CSD 2014), which is highly concentrated in the urbanised areas (average spatial density is 6690 persons per square km) (CSD 2014). Hong Kong GDP reached US$ 310 billion in 2015 (IMF 2015), which places the city at number five in the ranking of global cities (Hales et al. 2014). The territory of Hong Kong consists of three distinctive geographical areas: Hong Kong Island, Kowloon and the New Territories (see Fig. 6.1). Hong Kong Island and Kowloon Side encompass the most urbanised districts and more than 70% of the population of Hong Kong commute to these areas every work day (CSD 2014).

The public transit system of Hong Kong is a prominent feature of its socio-spatial organisation: 'Every day, over 12.5 million passenger journeys are made on a public transport system which includes railways, trams, buses, minibuses, taxis and ferries' (TD 2015). In addition, automobility by private car is another important traffic mode, especially for high earners given the onerous costs that it entails. By September 2017, there were more than 760,000 licensed vehicles in Hong Kong (TD 2017). Until the twenty-first century, bicycles were not considered as a transport mode by the local government and it was not a matter of concern in the sociopolitical debates about transportation. Although bicycles have nearly a hundred years of history in Hong Kong (Chu 2014, p. 56), for a long period they were just seen by transport authorities as a sport/leisure activity. By the end of the twentieth century, utilitarian cycling—cycling not just for sport/leisure purposes—accounted for about 0.5% of the daily weekday vehicle trips in Hong Kong. Furthermore, 97% of these trips took place in non-urban areas—mainly, the New Territories (TD 2004, p. 2).

Since the 2000s, however, some significant changes occurred. Cycling, both as sport/leisure and as a traffic mode, has become more and more visible in the public debate and was able to enter into the sociopolitical agenda. In addition to conducting several studies about cycling in Hong Kong (TD 2004, 2013), the government has also invested in building a more consistent cycle track network in non-urban areas

Fig. 6.1 Critical mass in Kowloon, Hong Kong, 2016 (*Source* Photograph by Miguel Angel Martínez López)

(LC 2014). The role of bicycles in Hong Kong's traffic system has been discussed repeatedly in the Legislative Council (LC 2011, 2013, 2016). Furthermore, there are more pro-cycling, non-governmental organisations emerging and active. On Facebook, by July 2016, there were at least 20 local groups that focus on various issues about cycling in Hong Kong. Cycling as a topic is also gaining more attention in the media, either through traditional outlets or through new independent media, following an international trend in other global cities (Aldred and Jungnickel 2014; Buehler et al. 2015; Fishman 2016).

All in all, it seems that cycling is a current sociopolitical concern in Hong Kong after a long period of being neglected in a city in which cycling used to be 'invisible and not available' (Zhao 2010) and where utilitarian cycling is still highly limited. And this kind of change is the consequence of the ongoing actions of and interactions between different relevant actors, especially between civic activists and state agencies. Nevertheless, the literature on this topic has failed to illuminate the sociopolitical dynamics behind the evolution of the cycling issue and the transport agenda. The aim of this article is thus to disclose the relations between cycling activists and state agencies, and how different civil society actors cooperate and/or compete to shape the legitimacy of alternative proposals regarding the existing policy, regulations and the built environment.

Cycling Governance, Cycling Field and Embedded Civic Activists

In the discipline of urban cycling governance, the issue of 'how to promote cycling' occupies an outstanding position. For instance, there are many studies focusing on how to change infrastructures (Parra et al. 2007; Pucher and Buehler 2009), built environments (Nkurunziza et al. 2012) and 'soft policies' (Jones and Novo de Azevedo 2013) in order to promote cycling in a certain city or country. Most of them focus on measures to promote cycling but they usually overlook the starting points of cycling promotion in a given city and simplify the contentious process by which cycling gains cultural and sociopolitical legitimacy. Following these critiques, a growing number of scholars identified some of the sociopolitical dimensions that bring about changes to urban cycling (Aldred 2013; Johnson 2011; Jones and Novo de Azevedo 2013). 'Who wants to promote cycling' and 'why do they want to' are now seen as the foundations of the 'how' question by many researchers. Sociopolitical agency and the context in which they operate have come to the fore of the research agenda on cycling governance.

Regarding the pro-cycling forces and their actions, two main kinds of organisations may be distinguished: non-governmental advocacy groups and governmental agencies. According to previous research, the inter-organisational and intra-organisational interactions among advocacy groups (Aldred 2013; Pucher et al. 2012), the dynamics within the government (Hulsmann 1997) and the conflictive interactions between non-governmental organisations and the government (Batterbury 2003; Xiong et al. 2012) are all key analytical dimensions. This approach may be supplemented with the conceptual tools provided by the 'organisational field' theory.

Bourdieu's notion of 'field' (Bourdieu and Wacquant 1992, p. 17: 'a field is a patterned system of objective forces [much in the manner of a magnetic field], a relational configuration endowed with a specific gravity which it imposes on all the objects and agents which enter it') constitutes the basis to study the institutional arrangements and social processes that influence organisational actions and 'situations where organised groups of actors gather and frame their actions vis-a-vis one another' (Fligstein 2001, p. 108). Some researchers treat the organisational field as predominantly static and emphasise the homogeneity of the organisations within a field. They focus on outcomes of field membership and institutional processes (Edelman 1992). Other scholars focus on the processes that create the homogeneity rather than the outcome itself (Hirsch 1997). Changes, variations and agencies have also been introduced into neo-institutionalist analyses (Seo and Creed 2002).

However, as Wooten and Hoffman (Wooten and Hoffman 2016, p. 10) pointed out, those approaches conceptualised field as 'things' that produced outcomes. Instead, fields can be conceptualised as 'mechanisms' (Davis and Marquis 2005) or 'relational spaces that provide an organisation with the opportunity to involve itself with other actors' (Wooten and Hoffman 2016, p. 11). We adhere to the latter and will now present the analytical implications for our research purposes.

First of all, the creation of the field and its evolution deserve an accurate description. 'Ideas play a pivotal role in motivating collective action, channelling policy resources, and shaping governance relations' (Bradford 2016, p. 659), but this dimension has seldom been applied to cycling governance. An exception is Horton's contributions. For him, the 'green' meaning of bicycles influenced both the emergent discourse and practice of a worldwide pro-cycling field (Horton 2006). In studying the rise and decline of Dutch cycling culture, Kuipers (2012) pointed out that the image of the bicycle, especially in the eyes of those in power, changes as time goes on and this certainly impacts on the evolution of the pro-cycling field.

Second, in a given field, there are organised or non-organised actors that can respond strategically to institutional pressures (Oliver 1991), lobby for the acceptance of new logics and practices, and engage in fighting for the legitimacy of the field (Lawrence 1999). They can be named as 'institutional entrepreneurs' (Wooten and Hoffman 2016, p. 9). This designation suits well civic cycling activists. Generally, they 'offer bicycle education and training, support funding and laws to improve bicycle conditions, and coordinate various pro-cycling events and activities' (Pucher et al. 2012, p. 337). To some extent, these actors/organisations are 'shaping the city from below' (Carlsson 2014) and build and maintain the legitimacy of urban cycling.

For instance, Aldred's research on how bike users in London promoted a 'pop-up' campaign to pressure the London mayoral candidates over cycling issues offers a good illustration of 'institutional entrepreneurs.' She argues that besides lobbying for cycling infrastructures, the campaign aimed at generating 'a positive cycling identity in the context of stigma' (Aldred 2013, p. 194). In a similar vein, other researchers concluded that pro-cycling forces paid more attention to achieve the legitimacy of biking by 'being' in the society instead of negotiating with the government (Blickstein and Hanson 2001). And the influences of multi-actor and multi-level participation in some relevant issues [e.g. climate policy making] are also explored by some existing studies (Setzer and Biderman 2013). Bike users have paid more attention to achieving the legitimacy of biking by 'being' in the society instead of negotiating with the government (Horton 2006). In many cities, collective events are organised by civic activists, but some of them reflect conflicts and even fights rather than negotiation between pro-cycling forces, anti-cycling ones, the police and the local government (Roth 2012; Shepard and Moore 2002)

The social science research about urban cycling in Hong Kong is still in an initial phase and there are limited existing studies. Several exceptions mainly focus on the government dimension, rather than civic activists. For instance, Zhao analysed Hong Kong governmental cycling policies and pointed out that under the 'scientific' surface there is a self-fulfilling prophecy within the governmental logic about urban cycling that put cycling and bicycle users in an unfavourable and marginal position (Zhao 2010). And in studying the utilitarian cycling in

Shatin, which is a non-central area in Hong Kong, Choi analysed not only governmental officials' views but also local residents' views and discovered gaps in their views on the potential role of bicycles and cycling (Choi 2007). Similarly, in studying cycling development in the New Territories in Hong Kong, Cheung explored the functions of local residents and cyclists in overcoming the weakness of governmental top-down planning processes to achieve sustainable mobility (Cheung 2011).

Despite the above contributions, it is not yet clear which are the different roles that field members or participants perform (Wooten and Hoffman 2016, p. 13), particularly when civic activists bring about certain changes. We assume there are significant dynamics rooted in both the connections among cycling activists and the interactions between them and other relevant actors, especially state authorities. In the following sections, we will draw on the above theoretical insights from the field approach in order to explain the cycling advocacy and activism issue in Hong Kong.

Research Method

In this research, we adopted a qualitative approach to address the research questions. First, we collected the existing documents and conducted textual analysis. The data about the government's traffic strategies from 1976 to 1999 are mainly obtained from the reports of three Comprehensive Transport Studies in Hong Kong, published in 1976, 1989 (1990) and 1999. They form the basis of traffic policies in this city, and the policy advice in each report provides the guidance for each following decade (Smith 1976; TB 1989; TD 1999). As for the new century, the data on the government's attitude and actions about cycling are obtained from various sources, including documents from the Legislative Council (Transport Panel) and Transport Panel Papers, supplemented with a series of transport reports and annual transport digests by the Transport Department (LC 2011, 2012, 2013, 2014, 2016; TD 2003, 2004, 2013, 2015).

Besides governmental documents, data from key pro-cycling activists' blogs, the websites of cycling organisations, and articles and editorials from newspapers and magazines were also collected. We select three cycling advocacy organisations as representatives: Cyclist Club, Hong Kong Cycling Alliance and Bike the Movement since they were established in different periods with different aims. We have reviewed all their announcements, activation records, policy reviews and presentations published on their websites and Facebook homepages. Furthermore, we have also reviewed the representative articles and editorials related to cycling in Hong Kong in eight widely circulated local newspapers[1] from 2004 to 2016, the period when cycling becomes more visible in the political agenda of this city.

Additionally, we also participated in regular cycling events and interviewed 19 relevant actors, held various focus group discussions and also engaged in private communications. The respondents include six leaders/core members of three local cycling advocating organisations/ groups, five officials from relevant governmental departments, three daily bicycle users, one district councillor, one local urban planning expert, one manager from a local traffic consultant firm and one local journalist. All the documents and transcripts were subject to a process of coding, analytical operations such as a timeline of events, identification of actors and interactions, meanings and attitudes attached to actions.

Findings

The Dynamics Rooted in Contextual Pressures

The issue of urban cycling in Hong Kong is attracting increasing public attention. Several conditions have contributed directly and indirectly to the rise of the pro-cycling issue. Two of these conditions seem to be the most significant.

First, other global cities have played a key role in boosting the legitimacy of cycling in Hong Kong by challenging the conventional paradigm in transport policies that ignores the potential of cycling as a traffic mode. In recent decades, under the worldwide accepted consensus of 'sustainable urban development,' automobile dependence has become one of the greatest challenges facing cities at the beginning of the

[1] These eight are selected from the 'List of newspapers in Hong Kong', accessed at https://zh.wikipedia.org/wiki/%E9%A6%99%E6%B8%AF%E5%A0%B1%E7%B4%99%E5%88%97%E8%A1%A8. They are *South China Morning Post, Apple Daily, Hong Kong Economic Times, Ming Pao Daily, Oriental Daily News, Hong Kong Economic Journal, Hong Kong Commercial Daily* and *Sing Pao Daily News*. For each newspaper/magazine, we chose the top 50 articles/editorials as ranked by 'relevance' when searching for '香港 + 單車' (Hong Kong + bicycle) on the website of each newspaper/journal.

twenty-first century (Balsas 2001). Cycling has been treated as an ideal mode of ecological transport in many international cities. In addition to the traditional bicycle-friendly cities in some countries, such as Denmark and the Netherlands (Kuipers 2012), numerous motorised cities in the UK, Germany, the USA and some developing countries have recently started to promote urban cycling (Pucher et al. 2010).

Hong Kong is a highly internationalised city, and changes in other global cities may be quickly echoed there. In fact, almost all public discussions about cycling in Hong Kong start with, 'London (or Paris, Singapore, and so on] has…'.[2] Two kinds of cities are always mentioned by cycling activists to criticise Hong Kong's situation. One category is other global cities, such as London, New York and Paris, which are perceived as ideal international examples for other global cities (Kuipers 2012). The second category encompasses Hong Kong's neighbouring and potential competitor cities, such as Taipei, Singapore, Guangzhou and some Japanese cities. They are usually mentioned when the feasibility of a given cycling policy in Hong Kong is discussed.

Second, the public's dissatisfaction with the prevailing and official 'Public Transit + Private Cars' transport system in Hong Kong has been in a continuous upswing, laying the sociopolitical foundation for vindicating cycling. Although the transport system in Hong Kong currently works efficiently (TD 2015) compared to major global cities,[3] its dominant dual structure is facing more and more problems. The number of private cars has soared rapidly in recent years, from nearly 381,000 in 2001 to 520,000 by 2015 (TD 2001, 2015). Traffic congestion has expanded to various urban areas. One interviewee described the following:

> In some parts, especially on holidays and weekends, serious traffic jams are emerging. Travel time (by car] is becoming more and more unpredictable. (HK_1, 25 November 2015)

[2] There are myriads of examples, such as the articles in some widely circulating local newspapers: *Ming Pao Daily* (http://www.e123.hk/ElderlyPro/details/297883). See also multiple links at the website and Facebook page of the Hong Kong Cycling Alliance (http://hkcyclingalliance.org/).

[3] For example, the average commuting time in Hong Kong is 29.2 minutes per day, and (only) 13% of citizens spend more than 45 minutes commuting per day, which is lower than in the UK (25%), Singapore (19%), and Japan (38%) (Regus 2011).

Although congestion primarily bothers wealthy car owners, the general public also has a significant array of complaints about the local transit system, such as over-crowded carriages, delays and other passengers' behaviour (Regus 2011). These problems are not unique to Hong Kong's transit system, but users here do not have many more alternative transport choices:

> Every morning, many of my friends need to spend eight to ten [HK] dollars to take the MTR [subway] to their workplaces and the same in the evening to return. The distance is not very far. Bicycles are a perfect alternative, but we cannot ride bicycles in the current traffic situations. (HK_6, 29 June 2016)
>
> I like the MTR and buses-they're good and necessary. We need them. But if you cannot afford [a car], they are your only choices, and that's not good. Why not give people more choices? If for some journeys I can use MTR, for some I can use bicycles, and for some I can walk. Then I think it will be much better. (HK_6, 17 May 2016)

Therefore, the lack of flexibility and diversity in the Hong Kong's transport system is at the root of the widespread cultural and sociopolitical endorsement of cycling promotion. Although these outside pressures and potential demands are influential triggers, they must still be elicited and brought about by local actors.

The Evolution of Cycling Activists and Their Interactions Aiming at Achieving Resonances with Other Actors

In the 1980s and 1990s, civic cycling activists and their organisations started to emerge in Hong Kong. At that time, most civic actors focused on sport and recreational cycling. In the new century, more utilitarian cycling advocacy actors have become active (Table 6.1). Discussions about the cycling issue also extend from cycling fans to the general public. In contrast to the cycling advocacy NGO in Guangzhou, for example, all utilitarian cycling advocacy organisations and groups emerging in the twenty-first century in Hong Kong are 'still based on volunteers, who are unpaid' (HK_19, 19 November 2016).

Besides formal organisations, some small groups and unorganised actors are also involved Hong Kong's cycling issue. The Internet, especially its social base, facilitates frequent and rapid interactions between

Table 6.1 Representative cycling advocacy organisations in Hong Kong

Organisations	*Year*	*Aim(s)*	*Main activities*
Cyclist Club	1985	'To promote cycling tourism in Hong Kong'	– Provide cycling safety information and skills – Weekly cycling activities in Hong Kong – Monthly cycling activities in mainland China
The Hong Kong Mountain Bike Association (HKMBA)	1994	'To promote trail advocacy through education, planning, funding, establishing and maintaining multi-use trails throughout Hong Kong'	– Organise mountain cycling activities – Provide information about mountain cycling in Hong Kong, along with trail guides, cycling skills and bicycle shops – Provide an online forum about mountain cycling in Hong Kong
Hong Kong Cycling Alliance	2003	'To promote increased levels of cycling in Hong Kong'	– Provide information about cycling in Hong Kong – Survey (questionnaire) – 'Ride of Silence' (annually, 2005 present) – 'Stand Up for Cycling'
Critical Mass HK	2004	'To raise awareness for cycling in Hong Kong'	– An unorganised event in which cyclists assemble and ride on the streets on the last Saturday of each month
South Island Road Cycling	2004	Road cycling and 'get home safely'	– Mid-week morning rides (5:45 am to 7:00 am) on the empty streets of Hong Kong Island
3 + 1 Cycling Club	2011	'Promote cycling culture'	– Foreign cycling trips – Urban cycling teaching – Regular ridings to request cycle tracks in the North District on Hong Kong Island
Slow-Mo Classic	2012	'To promote 'Slow Cycling' in Hong Kong'	– Provide (historical) information about cycling in Hong Kong – Organise cycling activities
Bike the Movement	2012	'To motivate people to bike more and love cycling in this bike-unfriendly city'	– Monthly bicycle gatherings – Provide bicycle travel stories and 'our bicycle and lifestyle store' – Hong Kong urban cycling life map (especially in urban areas)

these actors. More than 20 Facebook groups focus on cycling in Hong Kong. The leaders and members of cycling advocacy organisations, groups and individual actors interact frequently through these groups. In addition, there are also abundant offline face-to-face interactions and pro-cycling events. Of the regular urban rides, Critical Mass HK[4] stands out as the longest lasting.

A higher number of actors have participated in organisational and propagating activities and a specific form of institutionalisation has taken place. Thus, most organised and unorganised actors accept some basic beliefs, demands and strategies related to urban cycling. For example, the idea that 'cycling is a healthy and environmentally friendly travel mode' is emphasised by most cycling advocacy organisations' websites or Facebook homepages.

> Promoting legal and safe cycling and sharing useful information and knowledge about cycling in Hong Kong are key aims of almost all [pro-cycling] actors, although different people may have different interpretations of what is 'legal' and 'safe' cycling here. (HK_4, 17 May 2016)

Civic cycling activists also try to interact with other influential actors in the urban cycling field. For instance, some cycling advocacy organisations have regular meetings with governmental departments and officials, usually biannually, to exchange ideas and propose actions (HK_3, 6 January 2016). Representative institutions, such as the District Councils and the Legislative Council, and their members are key intermediary agents of civic actors in getting more sociopolitical support and interacting with the government. For example, in the HK Ride of Silence 2016,[5] district councillors and legislative councillors participated and gave speeches to support both the event and urban cycling, which attracted the attention of both the media and the government. Some of

[4]Critical Mass is a monthly ride with the purpose of raising awareness for cycling. It started in San Francisco in 1992 and has spread to hundreds of cities worldwide. In Hong Kong, it is held on the last Saturday of each month: https://groups.yahoo.com/neo/groups/massridehk/info, accessed 8 June 2016.

[5]A key annual event to remember injured cyclists in Hong Kong, which has lasted for 11 years: http://hkcyclingalliance.org/3167-2, accessed 8 June 2016.

these district councillors and legislative councillors were cycling activists, and the bicycle became a campaign symbol for some of them.

One example is the Legislative Council election in 2016. During this election, two candidates, one from a central urban area and one from a non-central area, clearly presented themselves as cyclists and cycling activists. They also proposed a pro-cycling vision for Hong Kong. Two interesting and significant points, however, must be pointed out. First, the bicycle, when used for electoral campaign purposes, enjoys more sociopolitical and even cultural meanings than just as a traffic mode. When these candidates discussed bicycles and urban cycling to the public and in the media, they referred not only to urban transport issues and environmental protection but also to more general sociopolitical issues, such as freedom, equality, justice and the 'free will of the Hong Kong people' (Chu 2016).

Second, both present candidates and prior pro-cycling candidates had close relationships with many civic cycling activists and their organisations, especially those involved with utilitarian cycling. Conversely, the civic actors used the election process itself as a sociopolitical opportunity to promote cycling. This explains why certain cycling advocacy groups and organisations actively helped these electoral candidates with various actions, such as sharing their Facebook posts on social media, donating to them and supporting the events they organised. Furthermore, both candidates won their elections. After that, one of them cooperated with a civic cycling advocacy organisation and organised a pro-cycling event in a central urban area, named Dawn-Ride (天光 Ride), which was planned to last for seven months.[6] They met at 7:00 am each Tuesday during this period and used bicycles to travel in the central area of Hong Kong Island.

The mass media are also used by civic actors to get more influential positions. In the 1980s and 1990s, there were few reports about cycling in the public media in Hong Kong. This has changed coming into the new century. Both the cycling issue and civic cycling activists have started to attract the media's attention. Interviews with cycling

[6]From 25 October 2016 to 11 April 2017. For more details, please refer to the website of this event: http://bikethemoment.com/tinkwongride/#comment-1953, accessed 25 January 2017.

activists were published in almost all of the important local newspapers,[7] and even some television programmes and independent films focused on the cycling issue in Hong Kong.[8] During these processes, civic cycling activists spread their understandings and visions of urban cycling in Hong Kong and criticised the Hong Kong government's decisions and attitudes towards cycling. By doing this, civic actors obtained some public support for both themselves and urban cycling in Hong Kong. They also received more legitimacy in engaging in the decision-making process regarding any relevant issue. As a result, some organisations, such as Cycling Alliance, have been invited to participate in some urban planning consulting processes.

Overall, the social relationships between civic cycling activists themselves and other actors have increased. Along with this process, some formal organisations, such as Cycling Alliance, have served as institutional entrepreneurs in the pro-cycling camp. Their practices and strategies to promote the use of urban bicycles have created patterns that have been gradually accepted by other cycling advocacy groups. This account would be superficial without considering the substantial differences between the cycling activists.

Divisions Among Cycling Activists and Their Reflective Interplay with the Government

By analysing the aims and demands set by various cycling advocacy organisations, it is easily determined that their ultimate goals have transformed from pure issue-based (distributional) goals to a hybrid of issue-based and identity-based (recognition) goals. Early organisations, such as the Cyclist Club and Mountain Bike Association, mainly focused on cycling tourism and mountain biking. For them, cycling was rarely

[7] Such as *Oriental Daily News* (http://orientaldaily.on.cc/cnt/lifestyle/20140116/00321_001.html), *Hong Kong Economic Journal* (http://forum.hkej.com/node/108454), *Hong Kong Commercial Daily* (http://www.hkcd.com.hk/content/2012-10/03/content_3057260.htm), *Sing Pao Daily News* (http://www.singpao.com/xw/ht/201309/t20130909_458237.html), and *South China Morning Post* (http://www.scmp.com/comment/insight-opinion/article/1885777/turn-hong-kongs-harbourfront-urban-haven-people-rather-cars).

[8] For instance, the independent film My Riding Diary directed by Wang Qin-yuan in 2009.

considered as an alternative traffic mode. However, the groups that came later, such as Cycling Alliance and Bike the Movement, shifted to cycling as a means of daily transport rather than just for leisure. Bike the Movement launched various actions urging the government to pay more attention to cycling but also tried to promote the public popularity of cycling and a sense of belonging for Hong Kong cyclists. Civic cycling activists can thus be divided into those who strive for recreational cycling along adequate and beautiful cycle paths and those who promote utilitarian cycling for urban commuting. Their interpretations of a bicycle-friendly city and the role of cycling in urban life differ substantially, although they are all helping to create a rich cycling advocacy field of social relationships and practices.

During the period when cycling was entirely ignored or marginalised in Hong Kong's sociopolitical agenda, the division among cycling advocacy entrepreneurs did not have any significant impact. At that time, the key issue was not how to promote cycling, but to make cycling visible—to make it matter. This suggests that the field was integrated, although not necessarily cohesive, due to the shared intention in awakening the sociopolitical awareness towards urban cycling. A representative of Cycling Alliance shared the following:

> My strategy is simply pushing out the messages to as many places as possible. To be quite honest, our focus has been engaging with the government, just to be there, with other organisations. We try to raise awareness amongst the government and the public and to engage the media in expressing the message of all aspects of cycling. There are various segments of cyclists. Consciously or unconsciously, we go together, make up a cycling circle community and get support from the rest of the community. (HK_3, 6 January 2016)

This organisations' activism paid off, as cycling became a more publicly debated topic. As the situation changed, the social relationships of those engaged in the cycling advocacy field and those from outside also started to change. In particular, the gap in the cycling advocacy field widened once the government started to react to the cycling advocacy field.

The Hong Kong government was basically absent in the cycling field before the 1990s. It then passively emerged as cycling increased in popularity and cycling advocacy forces grew. The government brought changes to civic cycling activists in interacting with them mainly after the release of the 'Two Worlds of Cycling' guideline, which implies a geographical division of territory where cycling is assigned differently (TD 2004). In the central urban areas, cycling is not encouraged (if not forbidden), whereas in the non-central urban areas, the government announced the building of a bicycle-friendly environment. The geographical separation led to the functional separation of cycling. This strategy also exacerbated the existing divisions in the cycling advocacy field.

As mentioned above, the cycling advocates were a mixture of those who predominantly perceived cycling as a healthy leisure activity and others who promoted it as an alternative traffic mode. Since the government announced the bicycle-friendly programme in the New Territories, the priority on the cycling agenda has been to build it up—for example, by dealing with the limitations of existing policies, infrastructure and safety training for cycling in this area. This was mainly conducted and inspired by recreational cycling activists and local cyclists in the New Territories. They accepted, positively or by omission, the Two Worlds of Cycling logic.

> It's a pity that this so-called 'cycling festival' just focuses on speed racing. We should tell people that cycling can be and is a part of life and our culture. (HK_2, 19 October 2015)

Some utilitarian cyclists launched an alternative Folk Bike Festival and called 'for all cyclists in Hong Kong to bring their bikes onto the streets close to their residences and ride along for no less than 30 minutes' (3 + 1 Cycling Club 2015). They claimed that 'the purpose is to demonstrate to the general public bicycles can be a carbon free, safe substitution for transportation in Hong Kong' (3 + 1 Cycling Club 2015). Most utilitarian-urban cycling groups immediately supported this folk festival. However, sport and recreational cyclists did not show much enthusiasm for the folk festival and even opined that it was causing trouble.[9] Supporters of each festival criticised each other in the media, in network forums and on Facebook.

Governmental actions promoting cycling were very limited. The conflicts between the two sides of the cycling advocacy field were partially

[9] More discussion about 'professional cyclists' attitudes can be attained at Fitz (Cycling), accessed at http://fitz.hk/sports/cycling.

due to the government's dominant view and its usual disregard for all dimensions of cycling. Therefore, this policy can be interpreted as a strategic reaction to the external and local pressures generated by the cycling advocacy field. Cycling became more socially and politically visible as a consequence, but the originally existing division in the field was exposed, expanded and exploited by the government's strategy: 'The government is happy to see it' (HK_3, 6 January 2016).

A Cycling Promotion Dilemma

'Everyone is dissatisfied with the situation' (HK_4, 17 May 2016) in the cycling advocacy field. Utilitarian cycling actors are the most dissatisfied. Although they now enjoy more opportunities to be involved in the political agenda, they still show deep discontent:

> We have lots of opportunities to express our views. But there is no further engagement and feedback. Whether anything we said influenced the process in any way-I have no reason to believe that. (HK_3, 6 January 2016)

What makes things worse for utilitarian cycling actors is that public debates about cycling are increasingly devoted to the issue of cycle tracks in non-central urban areas. In fact, little advancement of cycling policies in urban areas has been achieved since 2004, although the number of cyclists and activists is growing.

Recreational cycling actors are not satisfied either. The intended bicycle-friendly environment in the New Territories has not actually been friendly (Zhao 2010), as the construction plans for many cycle tracks have been repeatedly delayed, and some have even been cancelled.[10] However, cycling in urban areas is also becoming a demand for many recreational cyclists. Some cyclists wake up at 5 am or earlier to ride through the streets and roads for fun or for sportive purposes and go back home before the general public starts their commute. They then use another traffic mode to go to work because 'cycling in the urban area is dangerous and inconvenient' (HK_1, 25 November 2015).

[10]For more information: http://orientaldaily.on.cc/cnt/news/20150616/00176_035.html, accessed 8 June 2017.

Why do cycling activists have difficulty working together to push the government? Many questions have not been directly or collectively addressed by all of these groups. What should the role of cycling be in Hong Kong? What is the relationship between bicycles and other traffic modes? Why should cycling be promoted in Hong Kong? As cycling was officially ignored in the beginning, the different actors put these questions aside and strived to make cycling as visible as possible. The cycling advocacy field thus conformed to a low institutionalised relational space, within which members concentrated on short-term goals rather than on a shared long-term vision. When the short-term goals were partially achieved because of the government's strategic reactions to their pressures, the division within the camp was exposed and expanded.

The government is also unsatisfied with the current situation. Although its divide-and-conquer strategy has eliminated some of the pressure over the years, it must eventually confront the general outcry. First, utilitarian cycling cannot be entirely ignored, so the government occasionally carries out some promotional and educational advertisements. This eventually generates potential cyclists and growing public demands. Second, the creation of cycle tracks for recreational purposes must incorporate activists' and residents' concerns, thus delaying construction.

Overall, cycling in Hong Kong is not as developed as it is in other global cities. Nevertheless, a number of activist campaigns and cycling advocacy events have made the issue of cycling increasingly relevant. Different cycling advocacy groups initially succeeded in pressing the government and making the cycling issue visible. Although not many cycling policies have changed substantially over the years, transport authorities have reacted in very particular ways to the inescapable incorporation of cycling into the sociopolitical agenda.

However, cycling advocacy forces in Hong Kong are significantly diverse, which is a result of the history of cycling in this city. These forces cooperated during early stages when the government explicitly dismissed cycling. Their urgency to make cycling visible seemed to be a *sine qua non* condition to move forward. However, later on, their short-term goals prevented them from further cooperation and shared long-term views. Furthermore, transport authorities might have used the internal divisions within the cycling advocacy field as opportunities to continue with their low-intensity cycling policies. The above sections describe how the government, as a passive actor, reacted and coped with the activists' pressures. The Two Worlds of Cycling strategy reflects the sociopolitical differences of cycling advocates in rigid policies regarding the management of the urban territory and the role of bicycles in the traffic system. This did not prevent the authorities from dealing with even more demands from cycling advocacy forces. Likewise, the rising popularity of cycling has introduced new strains and contradictions that the conservative policies cannot address given the prevailing governance framework. Thus, the dynamic of social interactions has not yielded a significant advancement of cycling in Hong Kong and should be modified accordingly.

Conclusion

Coming into the new century, urban cycling has attracted increasing academic attention and become the new 'apple pie' (Spinney 2016, p. 451) in urban reseach. There have been a series of studies about certain actors whose interests, benefits and beliefs are related to the urban cycling issue and who have undertaken some kind of action towards it (Aldred 2012, 2013; Qian 2015; Yang et al. 2015; Zacharias 2002). Although some studies have paid their attentions to civic cycling activists' influences on this issue, they mainly treat the civic activists as a whole and analyse them as an individual actor who conducts rational actions to achieve their goals. The connections and potential divisions among these activists and between them and other actors, besides the interactions based on them, are relatively ignored. Therefore, this chapter focused on how cycling activists' emerged, evolved and strategically behave from a relational and dynamic approach.

The emergence and evolution of Hong Kong's cycling activists are embedded within the sociopolitical and particular transportation circumstances in this city. Other global cities' demonstration effect and the problems rooted in the dual-traffic system contribute to the emergence of the bicycles advocates. Through both online and offline frequent interactions, emerging cycling activists build both individual and organisational connections with each other. Besides, they also try to interact with other influential actors, especially the state authorities, in the urban cycling field in order to shaping the legitimacy of alternative cycling images and/or policy proposals. To some extent, they successfully transformed the issued from invisible to visible and are also becoming increasingly influential within the field.

However, as Hong Kong is a city in which bicycles are not widely used as a daily traffic mode by the main public, the cycling activists are a mixture of those who predominantly perceived cycling as a healthy leisure activity and others who fostered it as an alternative traffic mode. This entails an internal division between them on the understanding and vision of urban cycling in this city. That is why they are concentrated on short-term goals rather than on a shared long-term vision. When the short-term goals were partially achieved as a result of the government's reactions to pressures brought by activists, the division within the camp was exposed and expanded. As a consequence, this dynamic of social interactions has not yielded a significant advancement of cycling in Hong Kong.

References

3 + 1 Cycling Club. (2015). *The People's United of HK Celebrating Cycling Festival*. https://www.facebook.com/events/405858316286430/. Accessed 23 January 2016.

Aldred, R. (2012). The Role of Advocacy and Activism in Shaping Cycling Policy and Politics. In J. Parkin (Ed.), *Cycling and Sustainability* (pp. 83–108). Bingley: Emerald Group Publishing Limited.

Aldred, R. (2013). Who Are Londoners on Bikes and What Do They Want? Negotiating Identity and Issue Definition in a 'Pop-Up' Cycle Campaign. *Journal of Transport Geography, 30*, 194–201. https://doi.org/10.1016/j.jtrangeo.2013.01.005.

Aldred, R., & Jungnickel, K. (2014). Why Culture Matters for Transport Policy: The Case of Cycling in the UK. *Journal of Transport Geography, 34*, 78–87. https://doi.org/10.1016/j.jtrangeo.2013.11.004.

Balsas, C. (2001). Cities, Automobiles, and Sustainability. *Urban Affairs Review, 36*(3), 429–432.

Batterbury, S. (2003). Environmental Activism and Social Networks: Campaigning for Bicycles and Alternative Transport in West London. *The Annals of the American Academy of Political and Social Science, 590*(1), 150–169.

Blickstein, S., & Hanson, S. (2001). Critical Mass: Forging a Politics of Sustainable Mobility in the Information Age. *Transportation, 28*(4), 347–362. https://doi.org/10.1023/A:1011829701914.

Bourdieu, P., & Wacquant, L. (1992). *Invitation to a Reflexive Sociology*. Chicago, IL: University of Chicago Press.

Bradford, N. (2016). Ideas and Collaborative Governance: A Discursive Localism Approach. *Urban Affairs Review, 52*(5), 659–684. https://doi.org/10.1177/1078087415610011.

Buehler, R., Jung, W., & Hamre, A. (2015). Planning for Sustainable Transport in Germany and the USA: A Comparison of the Washington, DC and Stuttgart Regions. *International Planning Studies, 20*(3), 292–312. https://doi.org/10.1080/13563475.2014.989820.

Carlsson, C. (2014). *Shaping the City from Below*. http://www.boomcalifornia.com/2014/07/shaping-the-city-from-below/. Accessed 20 June 2016.

Cheung, C. (2011). *Planning for Sustainable Mobility: Implications for Cycling Development in Hong Kong*. Pok Fu Lam: The Unviersity of Hong Kong.

Choi, C. (2007). *Bicycle Use and Sustainable Transport in Hong Kong: A Case Study of Shatin*. Pok Fu Lam: The University of Hong Kong.

Chu, H. D. (2016) 朱凱廸單車政策倡議 [The Cycling Proposal of CHU Hoi Dick] (in Chinese). https://www.facebook.com/notes/八鄉朱凱廸-chu-hoi-dick/要自由-踩單車朱凱廸單車政策倡議/1097544766978113/. Accessed 25 January, 2017.

Chu, S. (2014). Bicycles in Hong Kong: One Hundred Years of History [香港百年单车简史] (in Chinese). In N. Tang (Ed.), *What. Issue 4:Free Riding* (pp. 56–79). Hong Kong: Joint Publishing (H.K.).

CSD (Census and Statistics Department). (2014). Number of Establishments and Persons Engaged (Other Than Those in the Civil Service) Analysed by Industry Division and District Council District. http://www.censtatd.gov.hk/hkstat/sub/sp452.jsp?productCode=D5250007. Accessed 11 October 2016.

Davis, G. F., & Marquis, C. (2005). Prospects for Organization Theory in the Early Twenty-First Century: Institutional Fields and Mechanisms. *Organization Science, 16*(4), 332–343. https://doi.org/10.1287/orsc.1050.0137.

Edelman, L. B. (1992). Legal Ambiguity and Symbolic Structures: Organizational Mediation of Civil Rights Law. *American Journal of Sociology, 97*(6), 1531–1576.

Fishman, E. (2016). Bikeshare: 'A Review of Recent Literature'. *Transport Reviews, 36*(1), 92–113. https://doi.org/10.1080/01441647.2015.1033036.

Fligstein, N. (2001). Social Skill and the Theory of Fields. *Sociological Theory, 19*(2).

Hales, M., Peterson, E., Mendoza Pena, A., & Gott, J. (2014). *2014 Global Cities Index and Emerging Cities Outlook*. https://www.atkearney.com/documents/10192/4461492/Global+Cities+Present+and+Future-GCI+2014.pdf/3628fd7d-70be-41bf-99d6-4c8eaf984cd5. Accessed 13 May 2016.

Hirsch, P. M. (1997). Sociology Without Social Structure: Neoinstitutional Theory Meets Brave New World. *American Journal of Sociology, 102*(6), 1702–1723.

Horton, D. (2006). Environmentalism and the Bicycle. *Environmental Politics, 15*(1), 41–58. https://doi.org/10.1080/09644010500418712.

Hulsmann, W. (1997). Towards the Bicucle-Friendly Town in Germany. In R. S. Tolley (Ed.), *The Greening of Urban Transport: Planning for Walking and Cycling in Western Cities* (2nd ed., pp. 287–298). Chichester: Wiley.

International Monetary Fund (IMF). (2015). *World Economic Outlook Database, April 2015.* http://www.imf.org/external/pubs/ft/weo/2015/01/weodata/weorept.aspx?pr.x=44&pr.y=20&sy=2015&ey=2015&scsm=1&ssd=1&sort=country&ds=.&br=1&c=532&s=NGDPD%252CNGDPDPC%252CPPPGDP%252CPPPPC&grp=0&a=. Accessed 4 June 2016.

Johnson, B. J. (2011). *A Case Study of American Bicycle Culture: How Cycling to Work Works in a Small Town in Kansas.* University of Kansas. https://kuscholarworks.ku.edu/bitstream/handle/1808/8790/Rodriguez_ku_0099M_11627_DATA_1.pdf;sequence=1.

Jones, T., & Novo de Azevedo, L. (2013). Economic, Social and Cultural Transformation and the Role of the Bicycle in Brazil. *Journal of Transport Geography, 30,* 208–219. https://doi.org/10.1016/j.jtrangeo.2013.02.005.

Kuipers, G. (2012). The Rise and Decline of National Habitus: Dutch Cycling Culture and the Shaping of National Similarity. *European, Journal of Social Theory, 16*(1), 17–35. https://doi.org/10.1177/1368431012437482.

Lawrence, T. B. (1999). Institutional Strategy. *Journal of Management, 25*(2), 161–188. https://doi.org/10.1016/S0149-2063(99)80008-7.

Legislative Council (LC). (2011). *LCQ9: Cycling Policy and Ancillary Facilities.* Accessed 11 September 2016.

Legislative Council. (2012). *LCQ2: 'Bicycle Friendly' Policy.* http://www.info.gov.hk/gia/general/201211/21/P201211210301.htm. Accessed 11 September 2016.

Legislative Council. (2013). *LCQ7: Bicycle Parking Spaces.* http://www.info.gov.hk/gia/general/201305/29/P201305290270.htm. Accessed 11 September 2016.

Legislative Council. (2014). *LCQ18: Development of Cycle Tracks.* http://www.info.gov.hk/gia/general/201402/12/P201402120245.htm. Accessed 11 September 2016.

Legislative Council. (2016). *LCQ19: 'Bicycle-Friendly' Environment.* Accessed 11 September 2016.

Nkurunziza, A., Zuidgeest, M., Brussel, M., & Van Maarseveen, M. (2012). Examining the Potential for Modal Change: Motivators and Barriers for Bicycle, Commuting in Dar-es-Salaam. *Transport Policy, 24,* 249–259. https://doi.org/10.1016/j.tranpol.2012.09.002.

Oliver, C. (1991). Strategic Responses to Institutional Processes. *The Academy of Management Review, 16*(1), 145–179. http://www.jstor.org/stable/258610.

Parra, D., Gomez, L., Pratt, M., Sarmiento, O. L., Mosquera, J., & Triche, E. (2007). Policy and Built Environment Changes in Bogota and Their

Importance in Health Promotion. *Indoor and Built Environment, 16*(4), 344–348. https://doi.org/10.1177/1420326X07080462.

Pucher, J., & Buehler, R. (2009). Integrating Bicycling and Public Transport in North America. *Journal of Public Transportation, 12*, 79–104. http://131.247.19.1/jpt/pdf/JPT12-3.pdf#page=82.

Pucher, J., Dill, J., & Handy, S. (2010). Infrastructure, Programs, and Policies to Increase Bicycling: An International Review. *Preventive Medicine, 50*, S106–S125. https://doi.org/10.1016/j.ypmed.2009.07.028.

Pucher, J., de Lanversin, E., Suzuki, T., & Whitelegg, J. (2012). Cycling in Megacities: London, Paris, New York,and Tokyo. In J. Pucher & R. Buehler (Ed.), *City Cycling* (pp. 319–397). Cambridge, MA and London: MIT Press.

Qian, J. (2015). No Right to the Street: Motorcycle Taxis, Discourse Production and the Regulation of Unruly Mobility. *Urban Studies, 52*(15), 2922–2947. https://doi.org/10.1177/0042098014539402.

Regus. (2011). 香港通勤的七宗罪 [Seven Problems About Commuting in Hong Kong] (in Chinese). http://press.regus.com/hong-kong/雷格斯調查報告羅列出香港通勤的七宗罪. Accessed 20 July 2016.

Roth, M. (2012). A Tumultuous Ride: New York Critical Mass and the Wrath of the NYPD. In C. Carlsson, L. Elliott, & A. Camarena (Ed.), *Shift Happens!: Ceitical Mass at 20* (pp. 235–254). San Francisco, CA: Full Enjoyment Books.

Seo, M.-G., & Creed, W. E. D. (2002). Institutional Contradictions, Praxis, and Institutional Change: A Dialectical Perspective. *The Academy of Management Review, 27*(2), 222–247.

Setzer, J., & Biderman, R. (2013). Increasing Participation in Climate Policy Implementation: A Case for Engaging SMEs from the Transport Sector in the City of Sao Paulo. *Environment and Planning C: Government and Policy, 31*(5), 806–821. https://doi.org/10.1068/c11262.

Shepard, B., & Moore, K. (2002). The Streets of New York. In C. Carlsson (Ed.), *Critical Mass: Bicycling's Defiant Celebration* (pp. 195–203). Oakland, CA: AK Press.

Smith, W. (1976). *Hong Kong Comprehensive Transport Study*. Hong Kong: Hong Kong Government.

Spinney, J. (2016). Fixing Mobility in the Neoliberal City: Cycling Policy and Practice in London as a Mode of Political-Economic and Biopolitical Governance. *Annals of the American Association of Geographers, 106*(2), 450–458. https://doi.org/10.1080/24694452.2015.1124016.

TB (Transport Branch of Government Secretariat). (1989). *Moving into the 21st Century: The Green Paper on Transport Policy in Hong Kong*. Hong Kong: Hong Kong Government.

TD (Transport Department). (1999). *Third Comprehensive Transport Study Final Report*, Hong Kong.

TD. (2001). *Vehicle Registration and Licensing Statistics*. http://www.td.gov.hk/filemanager/tc/content_535/table3_1.pdf. Accessed 10 June 2016.

TD. (2003). *Travel Characteristics Survey 2002-Final Report*. Hong Kong.

TD. (2004). Cycling Study Final Report. http://www.td.gov.hk/filemanager/en/publication/cyclingstudy.pdf. Accessed 16 July 2016.

TD. (2013). *Traffic and Transport Consultancy Study on Cycling Networks and Parking Facilities in Existing New Towns in Hong Kong*. http://www.td.gov.hk/filemanager/en/publication/td_194_2009_es_eng.pdf. Accessed 16 July 2016.

TD. (2015). *Transportation in Hong Kong*. http://www.td.gov.hk/en/transport_in_hong_kong/public_transport/introduction/index.html. Accessed 16 July 2016.

TD. (2017). *Key Statistics for September 2017*. http://www.td.gov.hk/en/transport_in_hong_kong/transport_figures/monthly_traffic_and_transport_digest/2017/201709/index.html. Accessed 13 December 2017.

Wooten, M., & Hoffman, A. J. (2016). *Organizational Fields Past, Present and Future* (Ross School of Business Working Paper No. 1311). http://papers.ssrn.com/sol3/papers.cfm?abstract_id=2767550.

Xiong, W., Chen, X., & Liu, Y. (2012). Public Participation inn Pedestrian and Bicycle Transportation Planning. *Urban Transport of China, 10*(1), 54–60.https://doi.org/10.13813/j.cn11-5141/u.2012.01.009.

Yang, J., Chen, J., Zhou, M., & Wang, Z. (2015). Major Issues for Biking Revival in Urban China. *Habitat International, 47*, 176–182. https://doi.org/10.1016/j.habitatint.2015.01.022.

Zacharias, J. (2002). Bicycle in Shanghai: Movement Patterns, Cyclist Attitudes and the Impact of Traffic Separation. *Transport Reviews, 22*(3), 309–322. https://doi.org/10.1080/01441640110103905.

Zhao, Z. (2010). 無法使用單車的城市 [The City in Which Cycling is Not Available] (in Chinese), http://www.ln.edu.hk/mcsln/23th_issue/feature_03.shtml. Accessed 5 April 2016.

CHAPTER 7

Urban Activism in Yogyakarta, Indonesia: Deprived and Discontented Citizens Demanding a More Just City

Sonia Roitman

Introduction

At the end of the past millennium, Indonesia embarked on one of the most important political transformations (policy decentralisation—*Reformasi*) that led to not only institutional changes at the national and local governments in relation to how policies are designed and implemented, but also social transformations encouraging community development, empowerment and participation. The country has had a strong economic performance over the last two decades, which has attracted more economic investment and has also expanded the purchasing power of middle-class groups who are developing new consumption practices. In addition, and in particular over the last five years, the country has strengthened its democratic system, electing as president for the first

S. Roitman (✉)
The University of Queensland, St Lucia, QLD, Australia
e-mail: s.roitman@uq.edu.au

N. M. Yip et al. (eds.),
Contested Cities and Urban Activism, The Contemporary City,
https://doi.org/10.1007/978-981-13-1730-9_7

time someone who did not belong to the military or political elite of the country.[1]

However, not all social groups have benefitted from economic development and policy decentralisation. Income inequality is increasing, and there is an unequal access to basic goods and services. Many citizens feel excluded from these changes and progress, but at the same time, they feel safer to express their opinions and demonstrate against unfair conditions. Several forms of activism have appeared in Indonesia. This article analyses two of these forms: one related to poverty alleviation and against social exclusion and a second one protesting against the commodification of the city and the idea of 'cities for people, not for profit' (Brenner et al. 2009). Those involved in these forms of activism challenge policies and practices without the intention of obtaining power for themselves in detriment of other groups (Martin 2007), but to make their voices, needs and demands heard and to create some change in their realities.

Through the analysis of these two types of activism, the article highlights the similarities and differences between these two types of collective action in Yogyakarta, Indonesia, including how they relate to institutional urban politics. These activities have been conducted by two groups of local residents. The first one is called Kalijawi and is formed mainly by women who live in informal settlements in Indonesia and have become organised to achieve better living conditions (especially in relation to land tenure). It is formed by those who are 'deprived of basic material and existing legal rights' (Marcuse 2009, p. 190). The second one is called Warga Berdaya and is formed by artists, environmental advocates and other middle-class residents and also informal settlement residents. They are not happy with urban public policies and the stronger role given to the private sector as producer of the city. It is formed by 'discontented' citizens, those who are 'discontented with life as they see it around them, perceived as limiting their own potentials for growth and creativity' (Marcuse 2009, p. 190). Even though these two groups are different, there are three similarities at the core of their practices. The first one is about the process and refers to the value of collective action as a tool to mobilise and create change. The second similarity refers to the

[1] Joko Widodo (known as Jokowi) is the 7th President of Indonesia, in office since October 2014. He was previously Governor of Jakarta (2012–2014) and Mayor of Surakarta (also known as Solo) (2005–2012).

overall aim consisting of a more just city and the realisation of an equal access to the 'right to the city'. The latter is translated into the type of use of the space and the process of participation in decision-making. Finally, a third one refers to the scale of their demands which is a local scale at the city and/or neighbourhood level.

This chapter is based on primary data collected through 15 semi-structured interviews, three focus-group discussions, attendance at presentations, informal conversations and regular contact with members of Kalijawi, Warga Berdaya and ArkomJogja[2] (the latter is the local NGO that supports Kalijawi) over a four-year period (2014–2018) as part of two research projects funded by the University of Queensland.

The article is organised into five sections including this Introduction. The next section discusses theoretical arguments related to the right to the city; then, the section, sociopolitical context after 1998, provides the sociopolitical context for the two case studies, while the section, Kalijawi and Warga Berdaya, goes into depth analysing Kalijawi and Warga Berdaya and finding similarities and differences between them. Finally, the last section highlights the value of this type of activism in Yogyakarta in relation to the international debate on social movements.

Demanding the Right to the City

Inequality is currently one of the most critical challenges faced by our societies (Alvaredo et al. 2017; Justino and Moore 2015; Oxfam 2015). The many faces of inequality show the inadequacy of the capitalist system, contributing to an immoral concentration of wealth as never before, the repression of women in some cultural systems that continue to undermine their role in society and the disparities that exist in terms of access to services, including health and education, which create gaps that are impossible to bridge, just to mention some of these faces. In this unequal environment, the demand for rights and the exercise of citizens' rights appear as refreshing and hopeful demands that, if realised, might contribute to reducing inequality gaps.

The 'right to the city', which is a concept that has been overly discussed in the academic debate in the last two decades, has appeared in the policy debate, even if timidly, and there are some groups who have taken it as one of their aspirations. Since its first discussion by Lefebvre in relation to the

[2] For more information on ArkomJogja visit www.arkomjogja.or.id.

'May 1968' movement, the concept has received different connotations. Lefebvre defines the right to the city as 'a cry and a demand. This right slowly meanders through the surprising detours of nostalgia and tourism, the return to the heart of the traditional city, and the call of existent or recently developed centralities' (Lefebvre 1996 [1967], p. 158).

Marcuse (2009, p. 190) further elaborates on the right to the city based on Lefebvre, arguing that this right is 'a cry out of necessity and a demand for something more' and explains:

> the demand is of those who are excluded, the cry is of those who are alienated; the demand is for the material necessities of life, the aspiration is for a broader right to what is necessary beyond the material to lead a satisfying life.

Further unpacking the concept, Marcuse asks himself 'whose right' are we discussing? And he answers: 'it is not everyone's right to the city with which we are concerned.... Some already have the right to the city' (Marcuse 2009, p. 191). Then, based on an analysis of class and culture, he identifies two social groups who struggle for the right to the city: the 'deprived' and the 'discontented'. The deprived includes two groups based on class: 'the excluded', who are at the margin of the system with no legal protection, and 'the working class', including unskilled and underpaid workers. These two groups are 'deprived of basic material and existing legal rights' (Marcuse 2009, p. 190). In terms of culture, he identifies those who are 'alienated', defined as

> of any economic class, many youth, artists, a significant part of the intelligentsia, in resistance to the dominant system as preventing adequate satisfaction of their human needs' (Marcuse 2009, p. 191).

These alienated citizens are those who are 'discontented with life as they see it around them, perceived as limiting their own potentials for growth and creativity' (Marcuse 2009, p. 190).

How do the deprived and the discontented citizens act to demand their right to the city? They become active citizens. The notion of citizenship escapes the traditional understanding of citizenship based on a social contract with the nation state to refer to 'new notions of citizenship' (Purcell 2003). These new notions still refer to rights, duties and membership of a political community (Purcell 2003), but the nation state loses weight and the international environment and the local level (city level) both get more relevance. Citizenship is thus being 'rescaled', 'reterritorialised' and 'reoriented' (Purcell 2003); the city becomes the new territory and the new scale where the demands for the exercise of rights are sought. Evidencing the multiple dimensions of the city and its complexity, the right to the city encompasses a long list of rights, including the right to housing, the 'right to stay' (or the right to secure tenancy), the right to safe water and sanitation, the right to affordable transport, the right to public space and so on.

Purcell (2003) suggests examining the right to the city according to two main rights: (1) the right to appropriate urban space and (2) the right to participate in the production of urban space. Appropriation refers to the right to use space, which is placed above the right of exchange and that the city fully meets residents' needs. Participation implies a central and more equal role of citizens in decision-making processes regarding the production of urban space. Hence, the right to the city challenges the capitalist system, giving priority to the use value of urban space over the exchange value and reinforcing that if all citizens have a right to participate, then property ownership no longer has a dominant voice in decision-making processes in the city (Purcell 2003, 2008). The right to participation also refers to citizens being able to influence, discuss and criticise the 'dynamics' and 'principles' of the spatialisation of social relations (Dikec 2001).

The right to the city refers to all residents (or inhabitants of the city, as mentioned by Lefebvre) and to the notion of inclusiveness. However, it has been criticised as putting more emphasis on the public sphere as a male-dominated space, without challenging gender divisions (Fenster 2011). Thus, the right to the city should include gender as an analytical category and should strive for women to be fully engaged in the public and private realms and able to participate in decision-making processes in the same form than men (Fenster 2011). Women are usually seen as part of marginalised groups either because they are poor or because they do not enjoy the same rights than men.

The deprived and the discontented citizens have started to engage in forms of active citizenship to demand the exercise of their right to the city. Citizenship is put in practice through activism, understood as an 'action on behalf of a cause, action that goes beyond what is conventional or routine' (Martin 2007, p. 20). Most activists pursue a change in the status quo, challenging policies and government practices. They seek to gain power through their actions, although power per se is not a goal. The goal is to change a particular situation. In the case of the 'right to the city' activists, they demand social justice in the city. These groups of deprived and discontented citizens do not usually deny the role of the state; on the contrary, they actively demand the government (at different levels) its responsibility to make possible their citizens' exercise to the right to the city and to change existing (unjust) situations. These citizens understand that in order for their rights to be valued, this has to be done by the government or at least started in partnership and dialogue with the government and followed by other stakeholders.

The right to the city involves practices and abilities (the right to vote, to participate, to make decisions) and also access to services and infrastructure at the neighbourhood and city level (access to housing, water, energy, social amenities, public space, public transport and safety) that leads to the enjoyment of the city. The right to the city is a collective demand that requires collective action. Citizens organise themselves based on their realisation that they need to generate power within, which is power built as a result of collective actions and shared beliefs to create change. As a collective right, the right to the city involves different groups, based on different affiliations, identity features and aspirations. Groups who are usually marginalised (the poor and women) constitute the core activists demanding this right, along with those who are 'discontented'. All these demands are taking place in different sociopolitical international contexts. In countries that have recently come back to democracy, like Indonesia, activists are regaining their rights to make their voices heard and participate in decisions affecting their lives.

Sociopolitical Context After 1998

Indonesia came back to democracy in 1998, when the 'Reformation Era' started, leaving behind the 'New Order', a period characterised by an authoritarian regime led by President Suharto in which all organised social groups who could potentially threat the government were dissolved. Citizens could not express freely and were discouraged to think critically. The new democratic era opened the doors for citizens to feel safe to express themselves, becoming involved in public issues and slowly regaining the value of collective action. This was encouraged with a strong decentralisation process started in 2001. This decentralisation involved the delegation of some powers from the national level to the provincial and local governments. It also promoted community participation of ordinary citizens, creating fora at the local level to steer discussions and for people to feel they could speak out, being their views considered. In this spirit, one of the most important programs for poverty alleviation (PNPM, National Program for Community Empowerment)[3] aimed at not only improving people's daily lives, through the provision of infrastructure and training to diversify livelihoods, but also encouraged the organisation of residents in community groups and their participation at the local level in making decisions on how to improve their neighbourhoods.

However, it has not been easy for people to start participating again. As one member of Warga Berdaya explained:

> actually the problem is not only about the room or space to give our opinion, but also the habits. If you have been oppressed for more than 30 years, you lose your capacity to talk. That kind of thing, we have to learn again, we have to learn how to express our anger. We have to learn how to express our concerns. People have to learn again how to express their opinion.

This statement evidences the heavy legacy of a long history of repression and authoritarian regime. It also shows how the discussion amongst activists is first linked to participation and democratic practices and later takes the form of more specific demands like the right to the city or the right to housing. This has been the case in other Asian countries. In South Korea, for example, in the 1960s and 1970s, activists demanded democracy, while in the 1980s social movements complained about real estate activities dominated by elite groups that evicted poor groups from prime locations and concentrated wealth in the elite (Shin 2017).

In these past 20 years, the Indonesian economy has performed strongly, increasing the GDP from USD$95 billion in 1998 to USD$932 billion in 2016. Poverty has decreased in relative numbers, from 23% of the population under the national poverty line in 1999 to 11% in 2014. Consequently, the middle class is expanding. The GNI per capita increased from USD$4,300 in 2000 to USD$11,220 in 2016 (World Bank 2017) and at the same time income inequality (measured

[3]The PNPM ran from 2007 to 2015. For more information on PNPM, especially in Yogyakarta, see Roitman (2016).

by the Gini coefficient) also increased from 0.31 in 1990 (UNDP 1990) to 0.41 in 2015 (World Bank 2017).

The continuing poverty situation of some citizens, the increasing income inequality, that is reflected in other types of inequality, and the emphasis on creating more democratic spaces for people to participate and demand for their voices to be heard are the conditions that have given room to new forms of activism in Indonesia. At the same time that there are more opportunities for democratic practices and participation and several groups have improved their living conditions, there are still some groups that feel excluded from economic progress and also from decision-making processes. Thus, they have organised themselves to show their discontent and marginalisation and their limited opportunities to feel equal citizens. The next section elaborates on the trajectories and actions by two of these new groups of activists in Yogyakarta. These groups have no political affiliation and no contact with political parties and, contrary to the situation in other countries, are not filling the gap of political opposition. On the contrary, in Indonesia, there are strong and antagonistic political parties. This would show a different situation from the democratic transition in Europe in the 1970s where social movements grew as alternative groups to the authoritarian regimes and decreased their role once political opposition consolidated (Pickvance 1995).

Kalijawi and Warga Berdaya

Over the last seven years, two groups of citizens have been active in showing their discontent and frustration with how Yogyakarta Metropolitan Area[4] is growing and some of the urban policies in place over recent years. They argue these policies are exclusionary and unjust and demand for equal right to the city and equal access to the city benefits. These two groups are Kalijawi and Warga Berdaya. Their origin, actions and achievements are discussed in the next two subsections. Later, a third subsection analyses the similarities and differences between the two groups.

[4]Yogyakarta Special Region (DYI) is a province of Indonesia located in Central Java (four million residents). It is a region and not a province because the head is the sultan (local king). Yogyakarta city is the capital of DYI (400,000 residents). The Metropolitan Area of Yogyakarta is the urban area, formed by Yogyakarta city and the urban areas of Sleman Regency and Bantul Regency (nearly two million residents). Jogja or Yogya is the short form for Yogyakarta.

Kalijawi

Kalijawi[5] is a community-based organisation or a network of informal settlement dwellers who live in Yogyakarta. It was created in 2012 and the majority of its members are women: '*women understand better their neighbourhoods*', said one of the members. There are around 250 households that belong to Kalijawi (KJ) and they live in 33 *kampung*[6] (informal settlements) located by the riverbank of Gajah Wong River and Winongo River (two of the three rivers that go across the city[7]).

KJ members are part of what Marcuse called 'the deprived'. They are citizens who cannot fully satisfy their material needs and are also in the margin of society regarding their legal rights. Additionally, most KJ members are part of the informal economy of the city. They have precarious jobs including market sellers, street vendors, cleaners, construction workers and scavengers. Their houses are small and built with simple materials. There is overcrowding, sharing places for resting, cooking, eating and also working on some productive activity. Not all houses have access to drinkable water and not all houses have a toilet. There are communal toilets in the neighbourhood. Most of these families have lived in the same house for years, even for generations. However, most KJ members do not have land certificates to prove land ownership. There are four main situations regarding tenancy: (1) land belongs to the government, which does not want residents to settle in the area as they are disaster-prone areas; (2) land belongs to private owners, who are usually 'absent landowners' as they do not live in the area or even in the city; (3) ownership is not even clear and disputes for land ownership amongst different parties are usual; (4) land belongs to the sultan of Yogyakarta and residents have permission to settle in those locations.

Similar to the situation in other cities and countries, in Yogyakarta *kampung* are heterogeneous settlements, with a diversity of income levels, housing conditions and forms of land tenure. KJ members are some

[5]The formal name is Paguyuban Kalijawi. *Paguyuban* means group in Javanese, *kali* means river, Ja refers to Gajah Wong River and Wi refers to Winongo River. These are the two rivers where Kalijawi members live. Kalijawi is then 'the group of the riverside people'.

[6]There are several meanings of *kampung*. In this case, they are considered informal settlements due to precarious land tenure and housing conditions. However, there is a variety of income and housing conditions and not all *kampung* residents are poor.

[7]Hereafter, city and Yogyakarta refer to Yogyakarta Metropolitan Area.

of the most vulnerable residents in each settlement, since they have low wages, inadequate housing and informal land tenure. This informality prevents KJ members from being able to improve their houses as they do not know how long they will be able to stay there and puts them in a situation of constant threats of eviction.

The main activities conducted by KJ over these five years of existence can be classified into three main groups: (a) co-production of knowledge on their situation; (b) generation of their own financial resources; (c) advocacy and establishment of relationships with other stakeholders, especially the local governments of Yogyakarta city and Sleman Regency, the two administrative areas of Yogyakarta Metropolitan Area where KJ members live.

The lack of information on informal settlements and the needs of their residents are a very common problem. The government is usually unaware of the scale of the problem, the severity of their lack of services and infrastructure and the informality of land tenure. To overcome this problem, in recent years community organisations in cities of the global South have carried out different types of exercises to collect data by themselves. Some of these exercises include profiles, enumerations, community mapping and poverty assessments. Enumerations are informal settlement censuses. Each house is visited to collect information about the household and the housing conditions. Profiles are descriptions of the main physical and social conditions in the settlement. Community mapping is the identification in the territory (translated into a map) of the resources, structures and services available in the settlement, the main problems and the resources and services required.[8] Kalijawi, supported by ArkomJogja,[9] has conducted community mapping exercises in 25 settlements. During the mapping, the community identifies their needs and problems, puts them on a map and also discusses possible solutions to these problems. It is a proactive co-production of knowledge that not only examines the problems, but more importantly, finds options for possible solutions that can be later discussed in dialogue with the government.

[8]For more information on profiles, enumerations and community mapping, see Patel and Baptist (2012).

[9]ArkomJogja is a local NGO that supports Kalijawi. For more information visit www.arkomjogja.or.id.

Kalijawi members suffer material deprivations. They are poor residents, with no regular income and low wages of around USD$110 per month, with just one or two people per household earning some wages.[10] Between December 2015 and January 2016, KJ undertook their first own poverty assessment. The process was facilitated by ArkomJogja and financially supported by UNDP. This assessment was part of a larger project supported by ACHR (Asian Coalition for Housing Rights)[11] claiming that it should be poor residents themselves who define their needs and the monetary amount required to cover these needs, and not 'experts', governments or academics who are not poor and do not understand and experience poverty on a daily basis.[12] This is part of a conceptual discussion started by Robert Chambers when discussing participatory approaches and the need for people to define their own needs, followed later by other researchers and activists and the criticisms of the Millennium Development Goals (Chambers 1995; Satterthwaite 2001, amongst several authors). The aim of this poverty assessment was to discuss the main dimensions of poverty, as experienced by poor residents themselves, rather than adopting the poverty indicators used by government and the United Nations.

Communities from five settlements where KJ has been working in collaboration with ArkomJogja participated in this poverty assessment,[13] constructing and deconstructing the meaning of poverty in a collaborative and bottom-up approach, emphasising the significance of the co-production of knowledge and the ownership of data by communities. Based on the community focus-group discussions for this assessment, four themes or dimensions appeared as critical everyday expenditures that should be considered when analysing poverty: land; housing and environment; security/vulnerability from disasters; and human capability (Table 7.1).

[10] This represents around USD$1300 per year, which is 10% of the average income for the country, as explained in the section 'sociopolitical context after 1998'.

[11] ACHR is an umbrella organisation that brings together informal settlement residents from Asian countries. For more information visit http://www.achr.net.

[12] For information on poverty assessments conducted by ACHR, see ACHR (2014).

[13] The assessment was done through a series of workshops and focus group discussions with five communities living in informal settlements in Yogyakarta Metropolitan Area. A structured questionnaire was also used to collect data from the participants. For more information, see Kusworo et al. (2016a).

Table 7.1 Poverty indicators by Kalijawi

Dimensions	*Indicators*	*Score*	*Weight*
Land	Insecurity of occupancy, threat from eviction	20	37
	Complicated land occupancy and/or tenure	17	
Housing and environment	People living in less than 9 m^2/person	3	10
	Renting or doubling-up a house with other people	3	
	Constructed from non-permanent and semi-permanent materials	2	
	Accessing electricity from neighbours	1	
	Lacking of sanitation facility: Private toilet	1	
Security from disaster	Insecurity from flooding	8	19
	Insecurity from landslide	5	
	Insecurity from fire	2	
	Insecurity from workplace eviction	4	
Human capability	Lacking of skill or unable to offer the skill	10	34
	Unable to cover children's education fee		
	Have many dependents	7	
	Less productive due to health and age	5	
	Unable to participate in the political process to influence decision-making	5	
		100	100

Source Adapted from Kuswuro et al. (2016b)

Considering income and the basic daily spending (mainly on food), monthly spending (on water, electricity, gas, transport, toiletries, house rent, health costs, etc.) and annual spending (to cover school fees, taxes, house renovation, land rent, social/religious events, etc.), communities established a minimum spending of IDR $960,000 per month per person (USD$71, around USD$280 for a four-member household and USD$2.4 per person per day). This is 2.5 times higher than the USD$110 per month per household that most Kalijawi families earn. This is also higher than the USD$ 1.9 per person per day currently considered by the World Bank as the threshold value for poverty. The participants of the poverty assessment were ranked according to the four dimensions of poverty and the indicators elaborated for each dimension. Based on their scores for these indicators, 58% were classified as 'very poor' (100-67 points), 30% were 'poor' (66-33 points) and 12% were considered 'not poor' (32-0 points) (Kusworo et al. 2016a).

Communities' main concern is their informal land tenure as it is the most difficult aspect to overcome due to the inability of KJ members to purchase the land where they are settled. Most KJ pay rent for the land where they are settled, pay land taxes to the local government and own their houses (structures). A minority group rent both land and housing structure. Land is expensive and an extremely limited resource in Yogyakarta and not available for sale in some cases (like government land). KJ members feel they can overcome other poverty dimensions (inadequate housing, vulnerability to natural disasters and limited livelihoods) but not land issues. They are not interested in getting ownership certificates, but they want the government to recognise and protect their settlements (Kusworo et al. 2016a). As long as they feel secure and they have 'the right to stay', land certificates are not important for them.

KJ members complain that while they are not allowed to live in the riverbank, there are several hotels that have recently been built there. Hence, they argue the government makes this difference because '*the hotel owner has "a bigger capital" and we don't have any capital*' (as expressed by KJ members). KJ members also feel frustrated when the government asks them why they live informally in the riverbank, without the government trying to understand their situation or without working together to find a good solution to their problem.

Their lack of secure land tenure creates two main problems for KJ members. First, they face a constant threat of eviction, despite being settled in the place for an average of 20 years. This has a negative impact on their emotional and physical well-being and also creates an obstacle for them to feel they can safely invest on improving their homes. Second, they have a complicated relationship with the government, which does not want to provide support for them as it disagrees on their location or does not want to legitimise their informal tenure, through government support. However, as explained later in this chapter, the relationship with the government is slowly changing due to the activities conducted by KJ.

One of the central demands of KJ is the 'right to stay', which is one of the rights included in the right to the city and is close to the right to appropriate space (Purcell 2003). KJ members want to be able to remain in their location, in a territory that they have been using and occupying for several years. The government argues that the riverbank is not an appropriate location for settling as it is an area prone to regular flooding. Communities have demand their right to participate in decision-making (Purcell 2003) as part of their right to the city. These negotiations include, for example, the extension of the setback from the riverbank. Using the information collected through the community mapping, and supported by Arkomjogja, KJ developed proposals for several *kampung* that have been submitted to the government. Some of them, like in the case of Kampung Mrican, have been accepted and others are still under negotiation with the government.

The second group of activities conducted by KJ are related to the generation of their own financial resources. This is done through the creation of saving groups formed of between 10 and 17 members who live in the same *kampung*. Each member contributes USD$0.15 daily. A common fund is created and can be used for community projects (construction of community toilets or a community centre) or be borrowed by members for individual projects (house improvement, livelihood activities, health or education fees). There are currently 23 saving groups and a community development fund (CDF) of USD$41,000.[14]

Establishing partnerships and relationships with other stakeholders is the third type of activity developed by KJ over these five years. This is done at three scales: local, national and international. At the local scale, KJ has made considerable efforts to establish a closer relationship with local governments through dialogue and through inviting government to participate in their activities and to understand their viewpoints. For example, some local government staff participated in the discussions when conducting the poverty assessment in 2015/2016. KJ also encourages relationships with other informal settlement residents and also other citizens of Yogyakarta. KJ organises an annual cultural festival by the riverside and invites all citizens to foster integration and awareness of living conditions and social relationships in informal settlements. This also helps to discourage prejudices against informal settlers. KJ members have also participated in events organised by Warga Berdaya.

At the national scale, KJ members participate in events related to housing and slum upgrading in other Indonesian cities and in exchanges of experiences with similar organisations based on other Indonesian cities. In 2016, for example, they participated in the Third Preparatory Meeting for Habitat III in Surabaya (Indonesia). This participation allowed them to attract the attention of the national government that was also present in the meeting.

At the international scale, KJ is part of ACHR since its creation and its members are in regular contact with similar organisations in other Asian

[14] Data for December 2016.

countries that are part of ACHR. They also participate in exchange visits to learn from other communities and share their experiences. For example, in July 2017, three members of KJ and ArkomJogja participated in an event in Mumbai organised by ACHR and SDI (Slum Dwellers International).

The main achievements of KJ can be considered as follows:

(a) Creation of the organisation which has become a solid group, emphasising the value of collective action;
(b) Consolidation and capacity building of the existing members, rather than expanding and growing as a group;
(c) Co-production of knowledge through community mapping and elaboration of solutions;
(d) Co-financing of communal and individual projects through group savings;
(e) Empowerment of women in a male-dominated society;
(f) Being able to establish dialogue with government and advocate for the needs and demands of *kampung* communities.

This last achievement is the most valued by KJ members as the government has become more open and receptive to their proposals and helps achieve a regular dialogue. However, each of the three local governments of the Yogyakarta Metropolitan Area has a different relationship with KJ (being more or less receptive to them). For KJ members, it is essential to be able to participate in decision-making processes on issues that will affect them. Thus, they feel they can exercise their right to the city in a more equal way. Women's empowerment is also an important outcome, as expressed by one of the ArkomJogja members:

> Kalijawi is an organization of cool and impressive women. Very great, they can think [and] find solutions for their own problems. They are superb, because women usually just follow others, follow their husbands. But in Kalijawi, women are the ones who produce the ideas and find solutions. It is not only their individual problems, but they also think about their neighbourhoods' problems. And then, they do savings, the purpose is in order to be independent and not... hoping [to get] external funding... They are an organization which tries to be independent, with [their own] savings and capacity building.

The demand for the 'right to stay' and the need to improve their living conditions are the main drivers of KJ members. Their main strategy to achieve these two goals is overcoming their exclusion as citizens through collective action and awareness. They want to be more visible in the city and want their demands to be heard by the government.

Warga Berdaya

Warga Berdaya is a group of citizens demanding of the local government better and fairer living conditions for all, with equal access to resources in the city. *Warga* means common people or citizens in Indonesian and *Berdaya* is empowerment. The members of this group are residents of Yogyakarta who face similar problems and have similar aspirations, especially in relation to the collective meaning of the city. It is formed by middle-class residents (artists, environmental advocates and students) and some low-income residents (some of them living in informal settlements). Its members are not happy with public policies at the city level and the stronger role given to the private sector as producer of the city. The group is formed by 'discontented' citizens (Marcuse 2009) who believe their access to use the city is limited, making their right to the city unachievable, constraining their opportunities for city enjoyment, while simultaneously commercial and residential developments benefiting just a few are allowed in the city. The right to the city refers here to the use and appropriation of space, as well as the participation in decision-making process, in particular when new projects will affect local residents.

Warga Berdaya (WB) started in 2010 with some people talking to each other about things they were 'discontented' about. It became stronger with its first public activities done in 2013, and it was at that time the name 'Warga Berdaya' was given to the group. There is no hierarchical structure in the group, with a core group of around 20 members, who regularly participate and design the actions to be implemented, and a larger group of supporters (sometimes larger than 100 people). WB does not receive any external funding and all activities are done with the members' own resources. These are not necessarily financial resources, but '*knowledge, and the capacity we have*', as expressed by a member. They define the group as 'an informal hub' that connects each other, trying to become '*empowered and critical citizens*'.

WB was born because some local artists, cyclists, researchers and students started to show their discontentment with how Yogyakarta was growing. There was a big increase in the hotel construction in the city[15] and, at the same time, public space was not looked after. There were no longer green areas for people to enjoy, with public squares converted into parking spaces and most public space occupied by commercial advertisements.[16] The improved purchase power of middle-class families had been translated in a higher number of vehicles in the streets, more traffic, vehicular chaos and collapsed road infrastructure. The group also claimed that the local government of Yogyakarta city (especially the mayor) was not doing anything to protect the 'city for all'. An interviewee said:

> one thing about this new [current] mayor and this new policy about Jogja and how Jogja is being handled recently that it opens too many doors for privatisation and also for commercialisation… if let's say that he wants [Jogja] to be more cosmopolitan then he does not offer a more solid vision or plan of how Jogja should be, instead just being a sell-out'.

WB thus criticises the policies ('principles') and 'dynamics' of real estate that contribute to injustice (Dikec 2001).

In February 2013, WB organised their first event called 'Merti Kutha' ('Clean city'). It consisted of cleaning some areas of the city centre from commercial advertisement and painting bike lanes in the streets. All was done overnight to avoid the police. Later in the same year, WB organised an art festival called 'Mencari Haryadi' ('Looking for Haryadi'), in reference to the mayor. The group was demanding his presence and an explicit position from his side on the role of public urban space. This art festival took place in a series of events from August to October 2013, including street painting (Yogyakarta is famous for its street murals) and street demonstrations outside the city hall with choir singing and parades. One of the artists interviewed commented that when the mural painting started in the city in the early 2000s, murals were mainly 'decorative', '*to make the city beautiful, to make the city different*', whereas since 2013 murals are '*a critic to the government*'. The government

[15] The number of hotels in Yogyakarta Metropolitan Area grew from 37 in 2005 to 85 in 2015. This does not include homestays or B&B.

[16] The cyclists were involved because the previous mayor (Herry Zudianto, in office 2001–2011) allowed cyclists to use the streets on Friday afternoons and evenings, closing the main street (Malioboro) to vehicles. People then used it for cycling, skating and walking. Haryadi Suyuti, the current mayor (in office since 2011), did not continue with this project and cyclists complained and demonstrated against this.

responded by declaring that painting murals was against the law if containing any sort of political slogans. However, artists continued to do it in a clandestine form.[17] Some WB members met with the mayor to try to find some solutions to their demands but they felt hopeless:

> ...the mayor has no vision. So, we have to blame ourselves because we chose him. This is the consequence of democracy. We voted the wrong person and we'll take the consequences for five years[18].

There was a 'Merti Kutha #2' event in May 2014. It consisted of cleaning heritage buildings from 'visual waste' (posters, pamphlets and graffiti) in the city centre.

The complaints against the commodification of the city became known as '*Jogja ora didol*' ('Jogja is not for sale'), opposing the undemanded construction of new hotels and other residential buildings in the city and the lack of public space for citizens to use and enjoy. WB members complained because even though there is a 'moratorium' to 'freeze' the construction of new hotels until the end of 2019, there were still new hotels under construction.[19] According to the local media, there were 104 new building permits submitted before the moratorium, with 80 of them under construction and 24 with pending approval on the basis of some further documentation still required.[20] The slogan 'Jogja is not for sale' became highly popular and was present in all demonstrations, media communication and forms of art.

Later in 2014, some low-income and lower-middle income settlements started to suffer a water shortage as a consequence of the

[17] In 2014, one student was put in jail for painting a mural related to 'Jogja is not for sale'. He had to pay a fine to get out from jail.

[18] Haryadi was re-elected and is currently in his second term in office (2011–2016 and 2016–2021).

[19] There are two 'moratoriums'. One in Yogyakarta city (Regulation 77/2013) halting the construction of new hotels from 1 January 2014 to 31 December 2016, later extended to 31 December 2017 (Regulation 55/2016) and another one in Sleman to stop development of new hotels, apartments and condominiums from 23 November 2015 to 31 December 2021 (Regulation 63/2015).

[20] *Source* http://jogja.antaranews.com/berita/336703/moratorium-pembangunan-hotel-diharapkan-berlanjut-di-diy, date accessed 5 July 2016.

development of new hotels that were using underground water that used to serve the nearby *kampung*. As part of the planning permit, new developments are obliged to get water supplied by the government water company and not from underground wells. However, several hotels were still taking underground water. This created conflicts between *kampung* residents who use underground water as their main water source and the new hotels as water access from wells diminished. A movement called '*Jogja Asat*' ('Jogja is drying out') was developed by *kampung* residents and WB to support *kampung* residents. As explained by a WB member, '*since 'Jogja is not for sale' became criminalised by the government, we changed it to 'Jogja asat' … It is always changing names, but the same people organised it. Same community, same activists*'.

Kampung residents demonstrated with banners in front of the new hotels taking underground water (August 2014), performing dances and other acts (like a sand bathing performance to make reference to the fact that residents could not bath with water any longer) to complain against the hotels and to make the problem visible and public. The most famous case was in *kampung* Miliran, where the construction of a new hotel (Fave Hotel) caused the *kampung* wells to dry since the hotel wells were deeper (80-metre-deep) than the *kampung* wells (40-metre-deep). As a consequence of the demonstrations and complaints from residents and WB, the local government intervened and closed the wells of the hotel, and water was recovered in the *kampung*. However, sometime later the hotel wells were reopened and the water shortage restarted.

In October 2014, the Kewek bridge located in a central area of the city was painted with a huge mural with images of fire, dragons, hotels, water and a giant representing 'urban development' eating other parts of the city and legends saying: 'Build wells not hotels', 'Jogja asat' and 'water = life'. WB made a documentary titled 'Jogja Asat' on the painting of the bridge. The information about the documentary was published in the website of IVAA (Indonesian Visual Art Archive), a local NGO in Yogyakarta that promotes local artists and supports WB, saying that the construction of hotels and shopping malls violates the rules of urban planning and denies the value of justice, increasing social disparities.[21] One interviewee expressed '*Yogya has become a victim of capitalism*'. Following Miliran *kampung*, other *kampung* in Yogyakarta also started

[21] *Source* http://ivaa-online.org/2015/04/28/video-warga-berdaya-jogja-asat/, date accessed 5 July 2016.

to complain about hotel development, the use of water and the threats of evictions. WB organised meetings in several *kampung* to make residents aware of these problems and organise possible complaint actions.

Warga Berdaya and Watchdoc produced another documentary in November 2014, launched in 2015, called '*Belakang Hotel*' ('Behind the hotel'). It is about the struggle of the residents of two *kampung*: Miliran and Gowongan. In the video, the leader from Miliran said that based on the Indonesian Constitution, land, water and natural resources in general should be controlled by the state and used for the welfare of people, not Fave Hotel (one of the hotels criticised in the documentary). A similar problem involved the residents of *kampung* Gowongan and Hotel 101, located in the city centre of Yogyakarta. The documentary mentioned that in 2003, there were 7,237 hotel rooms in the city, increasing to 10,303 rooms in 2013. The documentary was shown first at one of the local universities with 500 people attending the event. Later, it was shown in other venues, including some *kampung* to raise awareness of these problems.[22] WB has also held meetings with local government officials to have some discussions about new residential and commercial development affecting local residents. WB thinks that in most cases the government has not been open to their views and discussions have not resulted in positive outcomes.

WB is not against tourism. Yogyakarta is the second tourist destination in Indonesia after Bali and residents are aware of the benefits of tourism, but they think that the city does not need to build new hotels as there is not an increasing demand and the occupancy rate is low. They also want to promote small locally owned hotels and homestays, which, unlike large four and five-star hotels, contribute to the local economy. According to WB, hotel development has several negative impacts. First, a negative environmental impact as it uses underground water, when this is already a scarce resource. In this sense, WB says these new developments should comply with the required Environmental Impact Assessments before planning approvals are granted and that companies

[22]The documentary is also available on YouTube at https://www.youtube.com/watch?v=u8MhD3iy4rs&t=1296s (trailer and full movie, with English subtitles, date accessed 22 December 2017).

should comply with the requirement of using water from the network supply and not from underground wells. Second, it has negative social impacts encouraging strong divisions within local communities. Some hotels offered financial compensation to the outside residents to overcome conflicts. This created strong internal conflicts and divisions within each community as some residents accepted this compensation, whereas others have refused it considering it unfair and not what they want. The leader of Miliran confirmed in an interview the difficulties in mobilising all residents. Third, another negative social impact relates to strong rumours regarding hotel development been linked to corruption and money laundering. It is interesting that hotels in the city have a low occupancy rate of around 38%, and therefore, this creates questions on their financial survival.[23]

Since 2015, most of the activism by WB has been done through meetings with government and social media. A campaign was launched calling Yogyakarta residents to share problems related to 'Jogja is not for sale' through Tumblr, Facebook, Twitter and WB's blog. There is also a specific blog for people to discuss this issue.[24] People were requested to send pictures sharing information and complaints about new development projects in the city. Although it was not originally the intention, WB has become a repository of cases of conflicts between residents and the government (mainly about new commercial and residential developments). WB has also received strong media support and this has helped to spread out the information about their activities and actions. WB has then built a database and investigates these cases through the Internet and requesting public information on the permit process from the government. In 2015, WB and a legal NGO supported a group of residents who were against the construction of an upmarket high-rise building for students. They lost their legal case against the government.

In 2016, there was a discussion about the project for a new airport that will be built in a rural area of Yogyakarta Special Region (DYI). The discussion is about the land acquisition process, complaining that the local residents have not been consulted and not able to participate in the decision-making process. This was the first time WB's actions went

[23]The average room occupancy rate in 2016 was 43%, and just a bit higher in 2012 (46%) (DIY 2016).

[24]The blog is available at https://jogjaoradidol.wordpress.com/; it has not been very active since 2016.

out the city scale. Also in 2016, WB developed a program to support *kampung* residents affected by natural disasters through the actions of the Disaster Risk Forum, at the provincial level. This is a forum formed by members from the government, NGOs and private sector and managed by the NGOs. In 2017, WB was involved in public discussions on the role of civil society in local politics and city governance.

In these eight years of existence, the main achievements of WB have been as follows:

(a) Creation of an informal group that shows the value of collective action;
(b) Being able to make problems in the city more visible to public opinion and demand actions and responsibility from the government;
(c) Connecting groups of residents facing similar problems, creating awareness of these problems and supporting local residents;
(d) Channelling the voice of several heterogeneous groups feeling 'discontented'.

Despite these positive outcomes, some WB members feel that they have limited capacity to create change because they cannot change policies: '*We have power to say, but we don't have power to really change* [things], *because the change needs policy, and policy is made by the government*', one of the members expressed. In this regard, a WB member mentioned that one of their goals is to be able to change some of the local regulations about private development and investment in the city. This emphasises the strong role and responsibility the government has in building the city, even when change is initiated bottom-up.

For WB, art has a strong social function which is the ability to express injustices and they see art as the main vehicle to collectively express the movement: '*The silent people are the majority, the public is quiet, and artists are only a minority. But we have a voice, we have a tool, to speak, we have a tool to make a small voice become a big voice. With art and media*', explained one of the members.

Similarities and Differences Between Kalijawi and Warga Berdaya

Kalijawi and Warga Berdaya are groups that make claims for more equal access to the 'right to the city'. They have similarities and differences (Table 7.2). Both are recently created groups with a similar life

Table 7.2 Similarities and differences between Kalijawi and Warga Berdaya

Dimensions	*Kalijawi*	*Warga Berdaya*
Year of creation	2012	2010
City	Yogyakarta	Yogyakarta
Scale	Riverbanks *Kampung* (neighbourhood level)	City and localised (expanding to Yogyakarta Region)
Membership	Homogenous: Slum dwellers—'deprived'	Heterogeneous: Artists, environmental activists, students, slum dwellers—'discontented'
Overall goal	Right to the city and a more socially just city	Right to the city and a more socially just city
Actions/practices	Collective action Regular practices—Bottom up	Collective action Sporadic and specific practices—As a reaction—Bottom up
Type of activities	• Knowledge generation (profiles and community planning) • Creation of own financial resources (community savings) • Relationships with other stakeholders (including advocacy with government)	• Cultural activities (festivals, signing, documentaries, graffiti) • Street demonstrations • Residents' support, advocacy with government and knowledge generation
Visibility at city level	Growing, but still invisible for many	Higher visibility than KJ
Governance structure	Horizontal and formal	Horizontal, informal and loose
Relations with other stakeholders in Yogyakarta	Strong emphasis Local governments, universities, other city residents	Low emphasis Local governments, other city residents outside of WB.
Relations with other stakeholders outside Yogyakarta	Regular contact with other slum dwellers in other cities in Indonesia and in other countries	Limited contact

Source Elaborated by the author

trajectory. They both work in Yogyakarta; however, Kalijawi's scale is more specific as their members live and work on problems related to *kampung* located by Gajah Wong and Winongo Rivers. Warga Berdaya, on the other hand, works at the city level in any area and has recently gone outside the city boundaries in relation to the airport project in Kulon Progo Regency. In both cases, they work on specific projects, with very specific locations. They see urban transformation starting from small interventions and affecting change in the status quo through the cumulative effects of these small (but big) interventions.

The membership of both groups is different. In the case of KJ, members are a homogeneous group of residents living in informal settlements by the riverbanks who are deprived of the right to the city, whereas in WB, the membership is more heterogeneous who are discontented with how the city is used and appropriated and who can enjoy the city. Both groups share an ultimate goal of a more equal access to the city and a more socially just city. In neither of them, this goal is explicit, but underlies all actions undertaken. Unlike other social movements, these two groups do not put 'the right to the city' as one of their main slogans in their banners and informative material. Instead, 'the right to stay' and 'the city is not for sale' are used. Both groups advocate for bottom-up approaches for social change.

There are important differences between KJ and WB. The aims of KJ are more specific (to improve the living conditions in *kampung*, to get land tenure security, to get more diverse livelihoods), and therefore, there are regular activities done, including knowledge generation, creation of own financial resources and relationships with other stakeholders. For KJ, the advocacy and dialogue with the local governments are a central activity, understanding that they should work together with the government and not against the government. Both sides benefit from mutual cooperation, and it is important for both sides that *kampung* residents improve their living conditions. Conversely, WB's aims are more diverse and relate more to react against and confront the government for the outcomes of urban policies. To show their disagreement and discomfort, there are cultural activities, street demonstrations and residents' support and the generation of knowledge. One member of WB said: '*We have a bad and conflictive relationship with the government*'. However, in the last two years, WB has shifted from a confrontational approach towards the government to more advocacy and dialogue, even if they think the government does not listen to them. In relation to knowledge production, while there is an explicit awareness amongst KJ members on the power of knowledge and information, this is not the same in the case of WB where the creation of a database came as a consequence of other actions and not as a planned vehicle to gain power and negotiate with the government.

The relationships with other stakeholders have different value for both groups. WB has less interest in developing networks with other stakeholders in the city or outside the city as they do not seem to give strong emphasis to alliances with other actors, whereas for KJ this is seen as another central activity. KJ members have managed to establish strong relationships with other *kampung* residents in other Indonesian cities and other countries and also with local and foreign universities. In this spirit, KJ members have participated in events organised by WB and later WB thanked KJ for participating in their annual river celebration.

Although the network of relationships is broader for KJ than WB, the former has had higher visibility at the city level than the latter. This might be because the demands and complaints that are at the core of WB are more general (even if they also include claims by *kampung* residents) and can be understood more easily by Yogyakarta residents, regardless of social class and place of residence, while in the case of KJ, their demands are specific to *kampung* residents and therefore seem more distant to the needs and demands of other social groups in the city. Additionally, WB has received strong support from the media and even uses the media as a strong vehicle for internal and external communication. This is not the case for KJ.

Despite both groups having a horizontal governance structure, they have a different type of organisation. KJ works in a formal arrangement with organised committees for several activities and several members involved in each committee. There is also a leading group (even when all decisions are taken in meetings democratically according to what the majority wants). In the case of WB, there is no formal organisation, with small groupings being formed for specific activities and then disappearing to form new subgroupings or to have no subgroupings at all. WB wants to blend with 'the people', with 'communities': '*we are the people, we are the warga* [community]' and WB members are not interested in showing leadership. This might have future consequences in their work and limit their work.

Conclusion

Kalijawi and Warga Berdaya are two groups of activists produced by both unfair structural conditions (poverty, income inequality and an unequal access to use and enjoy the city) and the new democratic times in Indonesia. Through the analysis of the practices of these groups, this chapter contributes to the international discussion on social movements and activism. These social actors have been born as a consequence of the political changes in Indonesia where residents are re-learning how to freely express their opinions, claims and desires after many years of social and political repression.

These groups have different forms of working and expressing themselves showing 'new forms and practices of citizenship' (Purcell 2003, p. 564). They are formed by 'deprived' and 'discontented' citizens (Marcuse 2009) who do not have individual political ambitions but just seek transforming the status quo through concrete actions, in specific and small scales, being conscious of the need to strengthen their social capital and power within to transform their realities. Both groups show how the overall aim of 'the right to the city' for all is translated into 'the right to stay', the right to participate in decision-making processes and the right to appropriate and use urban space in equalitarian forms. These are claims and desires that are more concrete, affordable and accessible than the demands of past social movements in other regions. The struggles of these residents happen at the local scale of the city and the neighbourhood emphasising the value of everyday practices affected by these small actions and small change. The multiplication of these small actions at the local scale will lead to transformations at larger scales.

The actions of these two groups show the slow but positive sociopolitical change in Indonesia after years of repression that not only creates positive outcomes such as housing improvement and demonstrations against the commodification of the city, but also deeper transformations with citizens becoming more confident to speak out their needs and claiming for a collective and open process to create change. Groups formerly marginalised in the Indonesian culture (women, informal settlers and artists) are getting more power. However, it is interesting that neither of these groups has an explicit and upfront discussion about the meaning and process of construction of power, as was the case with social movements in the 1960s and 1970s in other countries. There is an emphasis on no hierarchies and no strong leaderships. At the same time, the construction of a relationship with the government through advocacy and regular dialogue appears as central to gain not only power but visibility and inclusion. Kalijawi and Warga Berdaya see the role of the state as critical in the production of urban space and urban society and in permanent dialogue and collaboration with civil society organisations. The state cannot and should not delegate its responsibilities, but needs to give space and equal opportunities to social groups working together towards more inclusive cities and societies where the equal access to the right to the city becomes a reality for all.

Acknowledgements I am deeply grateful to the interviewees who shared their views and experiences with me. Without their generous time and willingness to talk about their activities, this article would not be possible. Although this article expresses different views and perceptions, it remains my total responsibility as the author. My thanks also to Jason MacLeod, Shehana Gomez and the book editors who provided helpful comments on the drafts of this article.

This chapter is based on data collected for two research projects funded by The University of Queensland: 'Urban governance and housing policies in Indonesia', Sonia Roitman (Chief Investigator), UQ New Staff Grant (2013–2015) and 'How can gated communities contribute to the public good and improve the living conditions of poor residents? Gated communities and inequality in Indonesia', Sonia Roitman (Chief Investigator), UQECR Scheme Grant (2016–2018).

References

ACHR. (2014, September). Housing by People in Asia. *ACHR Newsletter*, No. 19.

Alvaredo, F., Chancel, L., Piketty, T., Saez, E., & Zucman, G. (2017). *World Inequality Report 2018*. The World Inequality Lab. Available at http://wir2018.wid.world.

Brenner, N., et al. (2009). Cities for People, Not for Profit. Introduction. *City, 13*(2–3), 176–184.

Chambers, R. (1995). Poverty and Livelihoods: Whose Reality Counts? *Environment and Urbanization, 7*(1), 173–204.

Daerah Istimewa Yogyakarta (DIY). (2016). *Room Occupancy Rate of Hotels 2016*, Badan Pusat Statistik. Yogyakarta: DIY.

Dikec, M. (2001). Justice and the Spatial Imagination. *Environment and Planning A, 33*, 1785–1805.

Fenster, T. (2011). The Right to the City and Gendered Everyday Life. In A. Sugranyes & C. Mathivet (Eds.), *Cities for All. Proposals and Experiences*

Towards the Right to the City (pp. 65–78). Santiago de Chile: Habitat International Coalition.

Justino, P., & Moore, M. (2015). *Inequality: Trends, Harms, and New Agendas* (Evidence Report No. 144). London: IDS.

Kusworo, Y., et al. (2016a). *Final Report: Participatory Study on Multidimensional Poverty in Yogyakarta* (Unpublished document). ArkomJogja.

Kusworo, Y., et al. (2016b). *Research on Participatory Multidimensional Poverty Assessment. Fact Sheet*, ArkomJogja.

Lefebvre, H. (1996 [1967]). The Right to the City. In E. Kofman & E. Lebas (Eds.), *Writings on Cities* (pp. 63–184). London: Blackwell.

Marcuse, P. (2009). From Critical Urban Theory to the Right to the City. *City, 13*(2–3), 185–197.

Martin, B. (2007). Activism, Social and Political. In G. Anderson & K. Herr (Eds.), *Encyclopedia of Activism and Social Justice* (pp. 20–27). Thousand Oaks: Sage.

Oxfam. (2015, January). *Wealth: Having It All and Wanting More.* Oxfam Issue Briefing. London: Oxfam.

Patel, S., & Baptist, C. (2012). Editorial: Documenting by the Undocumented. *Environment and Urbanization, 24*(1), 3–12.

Pickvance, C. (1995). Where Have Urban Movements Gone? In C. Hadjimichalis & D. Sadler (Eds.), *Europe at the Margins: New Mosaics of Inequality* (pp. 197–217). London: Wiley.

Purcell, M. (2003). Citizenship and the Right to the Global City: Reimagining the Capitalist World Order. *International Journal of Urban and Regional Research, 27*(3), 564–590.

Purcell, M. (2008). *Recapturing Democracy.* New York: Routledge.

Roitman, S. (2016). Between Top-Down and Bottom-Up Strategies for Housing and Poverty Alleviation in Indonesia: The PNPM Programme in Yogyakarta. In S. Attia, S. Shabka, Z. Shafik, & A. Aty (Eds.), *Dynamics and Resilience of Informal Areas: International Perspectives* (pp. 187–210). Cham: Springer.

Satterthwaite, D. (2001). Reducing Urban Poverty: Constraints on Effectiveness of Aid Agencies and Development Banks and Some Suggestions for Change. *Environment and Urbanization, 13*(1), 137–157.

Shin, H. (2017). Urban Movements and the Genealogy of Urban Rights Discourses: The Case of Urban Protesters Against Redevelopment and Displacement in Seoul, South Korea. *Annals of the American Association of Geographers*, https://doi.org/10.1080/24694452.2017.1392844. Advanced Online Publication.

UNDP. (1990). *Human Development Report 1990.* New York and Oxford: UNDP and Oxford University Press.

World Bank. (2017). *Indonesia Country Profile*, https://data.worldbank.org/country/indonesia. Date Accessed 18 October 2017.

CHAPTER 8

From the Squatters' Movement to Housing Activism in Spain: Identities, Tactics and Political Orientation

Robert González

Introduction

The purpose of this chapter[1] is to compare two current urban movements involved in noteworthy social battles inside Spain on subjects such as access to housing and spaces for communal sociability. Both the squatters' movement and housing activism display similarities and confluences, though they stand as two different movements not only in their development and organisational processes, but also in their goals and leadership.

[1]This paper is a result of the research project 'El movimiento de okupación de viviendas y centros sociales en España y en Europa: contextos, ciclos, identidades e institucionalización', which took place from January 2012 to December 2014. In depth, semi-structured interviews were carried out with 30 activists from both movements in major Spanish cities and included direct observation and analysis of documents produced by the movements themselves.

R. González (✉)
Universidad Autónoma del Estado de Hidalgo, Pachuca, Mexico

N. M. Yip et al. (eds.),
Contested Cities and Urban Activism, The Contemporary City,
https://doi.org/10.1007/978-981-13-1730-9_8

Why is it important to compare these two forms of urban activism? There are some advantages in a comparative approach. Following Tarrow (1986), comparative research in urban activism can help to improve analysis and interpretation of these forms of collective action in so many ways. First, the analysis of two similar forms of activism, such as squatting and housing, could reveal whether the use of different tactics, identities and strategies also results in different outcomes. In second place, regarding the history of both movements from a comparative perspective, we could observe whether a similar demand (housing) attracts support from different social groups if it comes from squatting or from new housing activism. In addition, if we analyse these two movements from a historical perspective, we should detect what changes in the political opportunity structure (POS) are more explicative of the emergence of new housing activism. Comparing these two movements across time and different stages could sensitise us to a variety of outcomes that urban activism enjoys or suffers under changing social, economic and political conditions. Applying Tarrow's example (1986: 149) about repression to our cases, we can ask ourselves if, although being a constant in state responses to urban activism, its forms, its degree and if it's accompanied by reforms or not, could vary in function of the different activist practices, strategies, tactics or identities. Finally, comparison across time can provide us empirical evidence about the novelty that new housing activism—especially the Platform for People Affected by Mortgages (Plataformas de Afectados por las Hipotecas, PAH)—supposes with respect to squatting tactics, strategies and identities around the same claim of achievable housing for people.

In this chapter, I argue that squatting and housing are two different kinds of urban activism that emerged in Spain as a response to the neoliberal urban-renewal regimes and the lack of housing policies (García-Lamarca 2016). The differences between these two forms of activism will appear in their tactics, identities and political orientations. Thus, the comparison will be useful to characterise two different but complementary forms of contemporary urban activism when faced with urban neoliberalism. These differences have been seen as a bifurcation of squatting activism into radicals (squatters) and moderates (housing activists) (Debelle et al. 2018); however, I would like to demonstrate that they are two different movements, as opposed to divergent. An understanding of the differences could help us to explain whether the outcomes of these responses to neoliberal regimes are also different. In fact, the 15M or Indignados movement should be seen as a hybridisation of these and other urban social movements (Martínez 2016). This is a specific condition of the Spanish case that has not been studied much in the literature of squatting and housing activism. In fact, some authors consider that housing activism converged with the 15M movement. In particular, these authors argue that the anti-eviction campaign initially operated as a legitimising mechanism for post-15M squatting, while struggles for housing were given a higher priority in the political agenda at later stages (Martínez and García 2013).

In this chapter, we will observe how housing and squatting activism have converged in many ways. The first proof of convergence was the presence of some squatting activists in the housing movement's first stages, especially in collectives such us V de Vivienda. Second, there is a transformation of some of the more significant activists from squatting activism to housing struggles—this was the case in the local PAH platforms of Barcelona and Sabadell. However, it was the 15M movement that led to the most convergence between both activisms. At the time, there was increasing legitimation of the occupation of social centres and buildings with residential purposes due to the mutual exchanges between anti-crisis 15M activists and anti-speculation activism from squatting (Martínez and Garcia 2013, p. 88). This process of urban activism convergence implies a social legitimation of insurgent urban practices, specifically the blocking of evictions and the occupation of housing buildings to host victims of the debt crisis. Some of the PAH's repertories and campaigns, such us stop evictions (Stop Desahucios) or social work (Obra Social), connect—even with different identities—with the traditional objectives and practices of traditional squatting activism. These insurgent urban practices imply the inversion, hybridisation, reconfiguration and politicisation of urban public space in Spain.

In the first section, I will discuss squatting practices that came on the European scene in the 1970s—and which was establishing itself in Spain during the 1980s. Then, I will explain the genesis of pro-housing activism as well as the rise of the PAH during the late 2000s.

Finally, I will compare these two urban movements by placing them in the wider context of the struggles against global neoliberalism. As other authors have demonstrated, 'the urban has (re) emerged in recent years as ground zero for insurgent activism across the world' (García-Lamarca

2016, p. 1). In order to analyse with a comparative approach, the principal characteristics of squatting and housing activism in Spain provide us empirical evidence of the organisation of popular discontent with the neoliberal urban regime in the South of Europe. We must notice, as have other authors, that squatting and housing activism practices have the potential to disrupt the dominant production of space through upsetting the dynamics of capital accumulation driving urbanisation in Spain (García-Lamarca 2016). But, squatting was a more avant-garde and earlier response, while housing supposes an extension of insurgent urban practices to larger sectors of the population. In other words, in this chapter we will analyse their similarities and differences in terms of identity, relationships with institutions, goals, organisation, the concept of squatting and the social composition of their activists in order to demonstrate empirically that housing activism has achieved more social impact than squatting, despite an apparently more moderate discourse. We will also demonstrate that the practice of squatting has been more legitimated due to the changes in identities, strategies and tactics involving the emergence of new housing activism, especially since the eruption of the 15M movement in 2011. Finally, we will ponder the similarities, differences and confluences between squatting and housing activism, remarking that both have performed insurgent practices 'with emancipatory potential because they enact equality for those who have no part in the dominant system' (García-Lamarca 2016, p. 3).

Squatting and Housing Activism in Spain: Different Responses to the Neoliberal Urban Regime

Here, I introduce the theoretical basis in order to analyse squatting and housing activism from the perspectives of both social movements and urban studies. First, I'll characterise squatting in Spain as a social movement following the precedent literature on this theme. Second, I'll delve deeply into two important social movements theories, political opportunity structure (POS) and frame analysis, that give us important theoretical keys to explain both the evolution of the squatting movement and the emergence of housing as a different movement, with its own identity, tactics and political orientation. Finally, I will use the concept of insurgent urban practices to analyse the emancipatory potential of new housing activism.

The squatting of derelict spaces to serve as social centres or homes fits the main notions that define a social movement: conflict, challenge, change and collective action. Squatting practices have transcended the field of protests, thus developing into a series of discourses, repertoire of actions and organised ways which provide a shared cultural identity, deeply related to the rise of the new social movements in Europe (Calle 2004, p. 273).

Yet, what does squatting mean? First, to squat means to live in (or give a different use to) abandoned facilities without the owner's consent (Pruijt 2004, p. 35). We could add that this movement focuses on direct access to a very limited kind of urban property (housing and social space) and the protection of these assets for self-defence (Martínez 2002). However, there are diverse squatting practices that coexist next to each other. Following common sense, one may differentiate those satisfying a need for housing from those turned into squatted social centres (SCCs), where countercultural activities take place in a public but non-governmental place—outside the bureaucratic logic of the state and the commercial logic of the private sector.

The social context that facilitates squats to appear is the result of de-structuring processes of social networks generated by neoliberal globalisation via the means of labour and life precariousness and a growing risk of social exclusion among the population. Labour precariousness and the difficulties in having access to housing have been an everyday occurrence in Spain since the eighties, but they have increased as a consequence of the 2008 financial crisis. For example, the youth unemployment rate reached to 43.5% in 2011 (Moreno et al. 2012). These two factors in particular have been merciless for young people who see themselves caught in the paradox of the existence of a huge quantity of derelicts resulting from real estate speculation and their own economic incapacity to afford decent housing.

On the other hand, both the culture and leisure coming from the private sector are becoming more and more inaccessible—with many huge discotheques and multiplex cinemas spread at large scale and encouraging consumerism. Besides their logistical 'benefits, the SSCs' task is to rearticulate local social networks and recover a self-governance that is both social and vital, in a positive way (Calle 2004, p. 283).

Squatting in Spain definitely has a dialectical position. In a way, squats are an end in themselves: spaces recovered from a property system that is based on speculation and on their exchange value over their use value. Yet, squatting also stands for a way of making a global battle against the system come true.

Stages of Squatting and Housing Activism in Spain

It is possible to split the history of Spain's squatting and housing activism into five stages. In order to determine where each of these stages starts and ends, we must take into account that their foundations are built on the protest cycles approach (Tarrow 1997) and on the changes of the squatting movements in relation to political opportunity structures (POSs).

POSs can be described according to six dimensions upon which political opportunities may vary. These include (a) the degree of openness of the institutional political system towards social movements; (b) the stability of the political elites' alignment; (c) alliances between movements and elites; (d) propensities towards the repression of movements; (e) the wider protest cycles at play; and (f) policies responding to the demands of movements (Brockett 1991; Diani 1998; Kitschelt 1996; McAdam 1998; Tejerina 1998). These structures from the political environment can either encourage or discourage collective action. This perspective could have problems, such as its deductive approach or the fact that it is sometimes more descriptive than analytical (Rootes 1999). However, I will embrace the POS approach to analyse what circumstances of the context have influence over the changes in tactics, strategies, identities and political orientation of the housing activism, in respect of the experiences of the squatting activism that preceded it.

According to Martínez (2018a), social movement scholars define POSs by highlighting the openness or access to public institutions, the cohesion of the elites, state repression, political alliances, media coverage and recognition, and, for some, also the subjective perception of those opportunities and constraints by the social groups involved. POSs help explain the different tactics and orientations adopted by the squatters (or by the emerging housing activists) in five consecutive stages covering the period 1977–2017. In particular, legislative changes led to the first-stage shift and the new forms of global mobilisations influenced the transition from the second to the third cycle, whereas the emergence of social movements at the national level was the most relevant context of the two final stages (Debelle et al. 2018). In fact, in these two final stages, it appears that housing activism would have converged with squatting despite the use of different tactics, strategies and identities.

Birth and Strengthening of Squatting (1977–1995)

Politically motivated squatting emerged in 1977 with environmental (Gallecs) and neighbourhood (Ateneu de Nou Barris) squatting but became consolidated in 1984 with the onset of the countercultural squats. These 1980s' occupations were mainly due to the search for spaces of freedom by young people who were followers of the punk ideology and its aesthetics. Soon afterwards, these squats assembled libertarian, autonomous and anarchist sectors. In Catalonia, the foremost squats were Ateneu de Cornellà (from 1986 to 2003) and Kasa de la Muntanya, squatted in 1989 (and still active today). In Madrid, an emerging Squatters' Assembly took part in the historical Minuesa squat in 1987. As for Euskadi (Basque Country), the most iconic gaztetxes (meaning 'houses for the youth') were those of Bolsa de Bilbao, Gaztetxe de Gazteiz (from Vitoria) and Euskal Jai from Pamplona. In 1992, there was a mutual influence between the squatting movement and other urban activists from the student, anti-war and feminist movements.

Golden Age (1996–2000)

The criminalisation of squatting in accordance with the new Penal Code (1996) marked the beginning of the expansion of squats. Furthermore, there was the movement's coming-out through the mass media due to the evictions of emblematic squats such as Cine Princesa in Barcelona and La Guindalera in Madrid. In this period, the squatters' movement became a role model for urban radical activism, carrying out demonstrations, resisting evictions and performing a large number of occupations. CSO Can Vies, in the Sants district of Barcelona, occupied in 1997, perfectly symbolises this period of movement growth (González 2015). However, towards 1999 and 2000, the squatter's inner circle was losing coordinated and organised spaces and the state's strategy of repression caused an enduring conflict with the police, reaching a critical point in 2001.

Some people who took part in the squatting protest were detained and charged with charges of terrorism in connection with armed groups (Asens 2004, p. 323).

The Squatters' Movement and Global Movements (2001–2005)

There are several contributions that led to the beginning of a new cycle within the squatter's movement starting in 2001 (Martínez 2007, p. 231). Changes in political opportunity structures were, among other things, caused by the launch of a new international protest cycle in 1999 (Seattle), which appeared in Spain in the 2000s. At the same time and in a sort of dialectics, squatters were also 'early risers' in the alterglobal movement, with some mobilisations before the 2000s (Martínez 2007). At this stage, the movement hybridises with other movements—anti-globalisation, resident activism and labour precariousness were all on the rise. SCCs such as El Laboratorio in Madrid or Can Masdeu in Barcelona stand out at this third juncture.

Housing and Squatting Bifurcation (2006–2010)

The squatting experience spread and went beyond its typical field of urban activism. Thus, for other subjectivities, squatting became useful as a powerful tool for resistance. Squats like Rimaia in Barcelona or Patio Maravillas in Madrid stand out as examples of this. On the other hand, a new social movement focused specifically on housing appears (V de Vivienda), which was supported by the moderate wing of the squatters' movement and represented by social centres like Espai Social Magdalenes (Fig. 8.1).

15M Movement and Its Effects on the Convergence Between Squatting and Housing Activism (2011–2017)

As a result of the 2008s global financial crisis, different housing movements, the 15M and cooperatives, squatted buildings in order to support people affected by a wave of widespread evictions from 2011 to 2013. Some authors point to it as the beginning of a new stage in the history of the squatters' movement, which turned into the generalised rise of squatting (Martínez 2018b). Conversely, in this period, through the 15M movement, squatting and housing activism converge with an 'accumulation of activists' exchanges' (Martínez and García 2015).

Fig. 8.1 Squatted Social Centre PSOA Malaya in Madrid (Spain), 2008 (*Source* Photograph by Miguel Angel Martínez López, permission to reproduce granted)

The New Housing Activism: Genesis and Characteristics

The housing problem and the economic crisis in Spain are the main contexts behind the rise of the pro-housing movement, led by the PAH (Platform for People affected by Mortgages). Due to the high prices of renting and buying a house, the crisis delayed the age of emancipation of young Spaniards, whose average is 29 years old. In fact, the percentage of emancipated young people aged 16–34 years, meaning that they dwell apart from their parents, was 44.1% in 2011 (Moreno et al. 2012, p. 180).

Average prices for buying and renting have been extremely high for several years. The huge number of empty apartments, resulting from real estate speculation, has done nothing but increase. A census from the Instituto Nacional de Estadística in 2011 counted 3.5 million unoccupied dwellings out of a total of 25 million, which supposes 14% (INE 2011). These numbers are due to a growing 'Spanish model', which is strongly based on homebuilding and which represented 9.3% of GDP by 2007 (over twice the figure in the USA) (Romero 2010, p. 18). Both the short-term nature of labour contracts and the high rates of unemployment (especially for young people) stand in the way of millions of individuals looking for housing. Turning housing into a mere commodity and the object of speculation has led to serious social, economic and environmental instabilities.

According to data from the PAH, an average of 500 people have lost their home every day since 2008. The foreclosure crisis has triggered evictions—over 600,000 from 2008 to 2014 (CGPJ, 2015 in García-Lamarca 2016)—along with the perpetual indebtedness of individuals (Colau and Alemany 2012, p. 32). At the same time, the absence of housing policies and the priority of the owner's interests and the market over the right to housing (although this is included in the Spanish Constitution, article 47) have become evident in the last few years. It has been caused both by the fact that 'prior to the financial crisis, 85 per cent of the population were homeowners' (García-Lamarca 2016, p. 8) and because of the marginalisation of the rental market and the virtual inexistence of social housing (Cabré and Módenes 2004).

The mistrust towards political parties, corruption, failed-to-fulfil promises and typical of European Union peripheral countries in crisis and under austerity (Alonso 2014, p. 21) complete the housing movement's social context. As in other southern European countries like Greece, the 2008 crisis led to an unexpected and deeper neoliberal turn on the part of the social democratic central government, bailing out many banks and cutting back fundamental public services (Martínez 2018b). The housing movement incorporates a large part of that democratic disaffection, which does not translate into political passivity in general, but rather into a greater preference for direct and non-representative participation models such as urban activism (Peña-López 2013). These conditions grew in 2011, when 'conservative and social democrats together prioritize[d] the satisfaction of creditors return in the national expenditure over other social needs' (Martínez 2018b, p. 41). These policy decisions, the general economic decline and the overwhelming number of political corruption cases undermined the legitimacy of the Spanish democratic regime among the citizenry, provoking the massive Indignados movement (15M) in May 2011 (Martínez 2018b; Castells 2012).

However, the housing movement emerged spontaneously a couple of years before the crisis due to an anonymous call on the Internet on 14 May 2006. The Occupy the Squares call was a huge success in several major cities, leading to unexpected demonstrations and assemblies. Activists from the squatters' movement, residents' movements and anti-capitalist leftists—along with a numerous group of 'newcomers' directly affected by housing unaffordability—showed up at all of these events (Fig. 8.2).

Fig. 8.2 Demonstration for the right to housing in Madrid (Spain), 2008 (*Source* Photograph by Miguel Angel Martínez López, permission to reproduced granted)

Activists provided tools for mobilisations while new 'militant' contingents brought freshness to them. For example, one of the most famous slogans or mottoes said, 'You won't have a house in your fucking lifetime' ('No tendrás una casa en la puta vida'), showing the popular masses' feeling of impotency and rage. Such displays would happen again at large on 15 May 2011 with the anti-austerity movement. Associations to the squatters' movement during this first stage are clear, by way of Espacio Social Magdalenes in Barcelona or Patio Maravillas in Madrid. Both of these squatted social centres agreed to negotiate with local governments, while the squatters' movement stood mostly against it. Espacio Social (ES) Magdalenes, squatted in May 2005, was supported by neighbours—victims of harassment from the landlord, who intended to build a hotel on the premises. Their skilful positive framing strategies towards the mass media[2] and their open willingness to negotiate made their continuity easier in spite of being located in downtown Barcelona. Nevertheless, the eviction on 11 April 2010 put an end to an intended dialogue that property owners did not listen to.

ES Magdalenes developed into a public project open to social movements and neighbours' associations. This space would promote cultural and political networks as well as initiatives articulating responses to the issues and social problems affecting Barcelona's historic city centre: growing gentrification, real estate harassment, pressure from the tourist industry, the undermining of social networks and shelter for the migrant population. In addition, ES Magdalenes was a meeting space for pioneering housing activism (V de Vivienda, Taller contra la Violencia Inmobiliaria y Urbanística) over the course of five years. With similar methods to ES Magdalenes, Espacio Polivalente Autogestionado (EPA) Patio Maravillas was squatted in Madrid in 2007. This place became crucial for Madrid's social movements because of its open identity and its architecture. The philosophy of Patio Maravillas was aimed at avoiding the squatters' stereotype, and it tried to create a socially and culturally participatory space whose openness attracted many social groups (migrants, local youth, artists, LGBTIQ people, etc.) and movement organisations. When facing eviction, their strategy was to gather social support and negotiate a possible expropriation of the facility with the

[2] See Máiz (1996) about framing strategies and political opportunities of social movements.

Council of Madrid. However, on 11 June 2015, Patio Maravillas was definitely evicted.

Having described the spaces for the first confluences or transitions between the squatters' and housing movements, let us go back to the latter's origins. After three multitudinous protests in 2006, a certain response by institutions—the Housing Act in Catalonia and national renting support—made the 2007 and 2008 calls look smaller. The Housing Act in Catalonia (Ley 18/2007, de 28 de diciembre del derecho a la vivienda) is a consequence of the consolidation of a social democratic government in that region and intended to incorporate some of the demands of the housing movements, such as fines for the big owners of empty buildings and actions in those in a poor state of conservation (art. 32 and 34 of Ll 18/2007). The national renting support policy of the Zapatero government (also with a social democratic stance) consisted in a lineal subvention for young tenants. Both policies were clearly insufficient, but they pacified the first wave of urban housing activism. However, the movement spread and gained in decentralisation and self-organisation.

The evolution in the foreclosure crisis and the 15M's upheavals would be the keys to the PAH's growth. In February 2009, in Barcelona, the first PAH centre was born as an initiative of V de Vivienda. In 2018, there were already 78 PAH centres in Catalonia and 245 around Spain (PAH 2018, http://afectadosporlahipoteca.com/contacto/#catalunya). Empowering tools developed by the people affected by foreclosures and the support of collective direct-action facing evictions introduced the PAH as a novel social movement. The above is clearly framed in what Pastor (2002) considers fundamental to the definition of a social movement: conflict, challenge, change and collective action. The PAH's repertoire of protest was combined with negotiating components involving administrations and financial entities as well as with boycott campaigns affecting the latter's public image.

The PAH has also made use of lawful and legislative means, introducing a citizen-initiated legislation (Iniciativa Legislativa Popular, ILP) for non-recourse debt with over a million and a half popular endorsements. In spite of being rejected by parliament, the process turned out to be in itself a substantive and symbolic impact of housing activism. Adell (2003) notes that the impact of social movements on the system can include recognition of the social movement as a legitimate actor (procedural or operative), a change or turn of the policy (substantive) or a transformation of the political context of the movement (structural). On the other hand, Ibarra et al. (2002) show that social movements could impact public policy in three dimensions.

1. The 'symbolic' or conceptual dimension that corresponds to the process of the construction of problems, specification of demands, elaboration of discourses supported by certain values, cognitive frameworks and belief systems, and, finally, the creation of public agendas for action. It is the dimension where the symbolic predominates the legitimating narrative of politics.
2. The 'substantive' dimension, which corresponds to the process of formulating policies and making decisions, that is to the phase where contents and options are negotiated and formalised through legally backed decisions. In the substantive dimension, interests and material resources come into play; objectives and models are set.
3. The 'operational' dimension, which corresponds to the implementation process. In it, mechanisms for the production of services, programs and projects are put into operation. Far from a technical conception of this dimension, new participatory and co-production spaces can be opened, linked to the management of resources, the evaluation of certain aspects and the consequent redesign of policies. Following these authors, this PAH initiative (ILP) causes a big operational impact (in terms of Adell) as well as a great symbolic impact, but finally it can't have a substantive impact (in terms of Ibarra et al.) because of the opposition of the right-wing parliamentary majority.

The PAH stopped 2045 evictions in eight years (2009–2017) (PAH, http://afectadosporlahipoteca.com/) showing its effectiveness as a mutual support mechanism and its negotiating skills. Furthermore, the PAH has rehoused over 2500 people through the PAH's obra social (social work), which consists of the occupation (known as 'recuperations') of empty bank-owned housing (García-Lamarca 2016).

Moreover, the PAH has a plural composition—mainly working-class migrants along with middle-class residents in a precarious or impoverished condition. As for age, the PAH displays a range that goes beyond young people.

The PAH agrees with the squatters' movement in defining injustice. However, one of the differences is that the PAH has introduced a series of reformist proposals. For the PAH, social usage should be prioritised over speculative usage, especially when it comes to empty apartments that belong to financial entities. The PAH supports social and affordable rent, considered a maximum of 30% of household income by housing activists. When facing emergency situations and in the absence of redistributive public policies, the PAH supports the squatting of empty buildings that belong to financial entities. At this point, squatting becomes a direct-action moving towards an end, but not an end in itself.

Finally, the leap taken into the political and institutional arena on behalf of the generation that drove movements such as 15M and the PAH came with widespread citizen candidacies for local governments in May 2015. For example, in Barcelona, the Barcelona en Comú platform, led by former PAH spokesperson Ada Colau, won the municipal election (Fig. 8.3).

Fig. 8.3 Protest action by the PAH in Barcelona (Spain), 2017 (*Source* Photograph by Miguel Angel Martínez López, permission to reproduced granted)

Squatting and Housing Activism: Similarities and Differences

Some authors have studied theoretical and practical differences between squatting and housing movements in other European countries. For Pruijt (2003, p. 145), one must draw a distinction between a pro-housing movement making use of squatting as a tactic and a squatters' movement to whom 'squatting itself is at the centre'.

According to Katz and Mayer (1983, p. 30), several structural variables within individual movements provide us with differences, which may match or collaborate on many occasions. First, the predominance of an autonomous ideology—which bears in mind that the making of antagonisms towards the establishment is the key to social change—is central in the squatters' movement. Moreover, the use of squatting for expressing and crafting counterculture becomes a distinct component of the squatters' movement against co-optation. Third, the squatters' movement always organises itself in a decentralised manner, while the pro-housing movement relies on more formalised structures as well as visible leaders. Finally, in the typical discussion about squatting as a means or as an end in itself, the squatters' movement recognises such ambivalence and sets its practice as a means to perform a broader social transformation and as an end in itself—because of the direct critique of private property (as a pillar of capitalism) and possibilities of creating social and vital autonomous islands in community centres and squatted houses.

Nevertheless, we cannot assume that squatters are not interested in housing. Moreover, it is a fact that housing activists have squatted buildings. But, what does the action of occupying mean to the PAH? It should be noted that this group has had a strong impact on the creation of counter-hegemonic spaces through the occupation of housing in order to accommodate evicted families. Thus, given the government's failure to deal with the real estate crisis, this social movement has built alternatives to the problem of evictions, re-housing 2500 people (PAH, http://afectadosporlahipoteca.com/). But what are the differences in the way this housing activism is approaching the action of occupying in comparison with the movement for squatting? What are the peculiarities of the first in relation to the second? Do these two movements have different identities and politicisation processes?

Two ideal models of pro-housing and squatters' movements might be useful to differentiate negotiation and confrontation strategies, as well as

Table 8.1 Differences between squatting and housing activism (ideal types)

	Squatting activism	*Housing activism*
Identity	Strong, countercultural	Diffuse, integrated
Relationships with institutions	Autonomous	Dialogue
Main strategy	Confrontation	Disruption/negotiation
Goals	Anti-capitalism	Housing policies
Organisation	Informal	Formalised
	Activists	Activists + people affected by foreclosures
Conception of squatting	End and means	Means to housing access
Age range	Youth mainly	Inter-generational

Source González (2015, p. 99)

the type of relationships with institutions, the goals, the organisational models or the very idea of squatting, as shown in Table 8.1.

The housing movement had an initial mobilising effect upon young people organising major protests on the right to housing in Spanish cities starting in 2006. Due to the 2008 crisis, this movement reached other social sectors such as those affected by foreclosures. This situation makes it clear that the squatters' movement is, essentially, not a pro-housing movement. In fact, it should be typified as a movement where political motivations that break from the capitalist system coexist with alternative strategies looking for housing and communal living spaces.

Furthermore, the radical nature of squatting proposals and their attack on private property have made it difficult to frame their discourse inside dominant hegemonic frameworks. Instead, the strong social support of the 15M movement and the PAH opens a new cycle of struggles centred on basic demands against the neoliberal model of urban capitalism and in favour of real democracy, autonomy from the market and punishment to the corruption of the political elite.

In Spain, public administrations have not been influenced yet by independent social movements. Indeed, recent trends show an increase in police and legal pressure on squats. These legal and police pressures are happening more and more, particularly due to reasons that are related to poverty. Thus, according to data from the Barcelona City Council, the number of squats expanded 11.2% in 2009, reaching a total of 249 in the city (Manrique 2010). These numbers display how the social, economic and urban conditions that gave rise to the squatting phenomenon have not diminished but increased. In such a context, the prospects for the squatters' movement seem uncertain. On the one hand, this movement is at the centre of a criminalisation spiral while, on the other hand, squatting practices have spread as never before to other social movements as well as to people having trouble with housing access. The main evidence of coinciding strategies between the squatters' and pro-housing movements are the very squats promoted by the PAH.

The majority of the interviewees justified squatting as a last resort: a means to solve the basic need for shelter in a context of rising unemployment and impoverishment. '(…), the occupation has changed a lot, okay. People now occupy by necessity, okay. The vast majority' (interview with J., PAH Sabadell, 2014). Squatting is thus framed as a way to recover housing for those with unbearable financial debts. The PAH squatters only occupy bank properties as a way of recovering what 'was already theirs'. They do not see themselves as okupas (squatters). For them, squatting has the aim of forcing the bank to negotiate affordable rental contracts. 'I think that social housing would be better. But if someone is willing to take care of (an unoccupied building], I do not know what I would do. A father with two children. If I had to occupy or stay in the street, I would' (interview with S., PAH Barcelona, 2014).

Some interviewees also express their support for SSCs if the buildings were abandoned and state-owned, while others reproduce and reject the stigmatised image of the okupa.

Interaction between the PAH and the squatting activists happened behind the scenes, and some specific actions show that both have cooperated with each other (Martínez and Garcia 2015). This cooperation is illustrated, for example, by the squatting of the 15-O building in Barcelona in 2012. Furthermore, in 2013, the PAH Sabadell occupied three buildings owned by the public bank SAREB and then negotiated a deal for the persons involved in the action; squatting activists also participated in these local PAH actions (Debelle et al. 2018).

The PAH has had a strong political impact and has attracted attention in recent years. In spite of its apparently moderate themed profile, the PAH is combining political tendencies for democratic regeneration and radical change in the political and economic systems. Moreover, the PAH is performing as a social movement, producing a collective consciousness, since it has been able to openly employ an injustice framework that has reached a significant portion of the population. Based on this scheme, the PAH has created a collective action framework, that is an individual positive disposition to carry out actions against evictions by people who were not activists in the past. The PAH's practice of blocking evictions is therefore seen as an insurgent practice that has an impact on the creation of urban space. Actually, by 2016, the 'PAH had successfully blocked over 2000 evictions across Spain' (García-Lamarca 2016, p. 9).

The request for new modes of democratic participation and the connection to a material right, such as housing, bring housing movements down to earth—more clearly in step with public opinion than the squatters' movement. The former vindications of the 'reformist left' place it closer to the symbolic frameworks of most citizens. The different framing strategies are located on a continuum that goes from the most demanding and transformative to the most adaptive, and in tune with the orientations and preferences of the mobilisation potentially available (Máiz 1996). While the former, the frame transforming (such as the squatting ones), seeks to radically modify the values and attitudes of the mobilisation potential, the second, the frame bridging, adapts its rhetorical strategy to the current state of opinion of the mobilisation potential. Given that both strategies are fully compatible, even from ambivalent use strategies (e.g. the PAH ones), it is expected that the second strategy has more short-term political impact and facilitates the activity of creating critical-promoting coalitions. On the other hand, a predominance of the transforming framing strategy will present more difficulties in impacting during the short and medium terms (González 2018). All of these features place housing activism as a crucial front in the protest cycles against the neoliberal management of financial crisis and as a result, bring it—from its very practice and paradoxically (or not)—closer to the squatters' movement's own anti-capitalistic origins.

Conclusions

The first conclusion addresses the goal of looking at whether the housing movement that emerged in the late 2000s can be considered a new stage in the history of squatting or not, and whether it must be approached as a separate and differentiated movement from that of the squatters'. This article proves that squatters' activism, or some parts of it, merge with the genesis of new housing activism.

Moreover, squatters were pioneers in fighting real estate speculation and neoliberal urbanism in Spain. After more than 30 years of existence, the squatters' activism, articulated around social centres, faced the issue of legalisation in some cases. The differentiation process that resulted from these splits shaped the discourse of the housing activism that rose to prominence from 2006—and more from 2009—onwards (Debelle et al. 2018).

Despite the former squatters' experiences having nurtured the new housing activism and having contributed to the new waves of both squatted and non-squatted independent social centres, following the mobilisations of the 15 M movement, it has been made clear as well that the differences between squatting and housing activism are significant in terms of analytical dimensions: the diffuse and integrated identity of the pro-housing movement in contrast to the stronger countercultural squatters' movement; an immediate willingness for negotiation from the pro-housing movement and a more autonomous strategy from the squatters' collectives; specific goals focused on housing policies on behalf of the former and some more general objectives which connect with the most radical and anti-capitalist traditions from squatters; a more formal organisation with known representatives, with differences between activists and those affected by foreclosures, in contrast to the high diversity of the squatters' organisation using different forms of coordination based on the everyday work of activists in the realm of squats. The wide scale and visibility of the PAH seem to indicate that squatting enjoys more legitimacy today. As a consequence, the PAH and its occupations are willing to engage in negotiations with the ruling classes.

Throughout the different cycles, the emergence of new movements destabilised the previously existing squatters' practices and then translated them into new forms of activism. However, due to the differences between squatter and housing activism practices, the housing movement cannot be taken as the fifth stage of squatting but a differentiated form of activism in and of itself.

Finally, both the pro-housing activism and the squatters' movement pose a social and cooperative economy taking population needs back to the centre and displacing profit-making from the centre of urban life. It is possible to assert that—even if making use of different tactics, strategies, identities and political orientations—the ultimate goal of both kinds of activism agrees on transforming the system, and beyond solving a public policy housing problem, its solution should be linked to a more global change in the political and economic system.

References

Adell, R. (2003). El estudio del contexto político a través de la protesta colectiva. La transición política española en la calle. In M. J. Funes & R. Adell (Eds.), *Movimientos sociales: cambio social y participación* (pp. 77–108). Madrid: UNED Ediciones.

Alonso, S. (2014). «Votas pero no eliges»: la democracia y la crisis de la deuda soberana en la Eurozona. *Recerca, Revista de Pensament i Anàlisi, 15*, 21–53.

Asens, J. (2004). La represión al movimiento de las okupaciones: del aparato policial a los mass media. In R. Adell & M. Martínez (Eds.), pp. 293–337.

Brockett, C. (1991). The Structure of Political Opportunities and Peasant Mobilisation in Central America. *Comparative Politics, 53*, 253–274.

Cabré, A., & Módenes, J. A. (2004). Homeownership and Social Inequality in Spain. In K. Kurz & H. P. Blossfeld (Eds.), *Home Ownership and Social Inequality in a Comparative Perspective* (Stanford, CA: Stanford University Press).

Calle, Á. (2004). Okupaciones: un movimiento contra las desigualdades materiales y expresivas. In F. Tezanos (Ed.), *Tendencias en desigualdad y exclusión* (pp. 270–305). Madrid: Sistema.

Castells, M. (2012). *Redes de indignación y esperanza: los movimientos sociales en la era de Internet*. Madrid: Alianza Editoral.

Colau, A., & Alemany, A. (2012). *Vidas hipotecadas. De la burbuja inmobiliaria al derecho a la vivienda*. Barcelona: Cuadrilatero de Libros.

Debelle, G., Catteneo, C., González, R., Barranco, O., & Llobet, M. (2018). Squatting Cycles in Barcelona: Identities, Repression and the Controversy of Institutionalisation. In M. Martínez (Ed.), *The Urban Politics of Squatters' Movements* (pp. 51–73). New York: Palgrave Macmillan.

Diani, M. (1998). Las redes de los movimientos: una perspectiva de análisis. In P. Ibarra & B. Tejerina (Eds.), *Los movimientos sociales, transformaciones políticas y cambio cultural*. Madrid: Trotta.

García-Lamarca, M. (2016). From Occupying Plazas to Recuperating Housing: Insurgent Practices in Spain. *International Journal of Urban and Regional Research*. https://doi.org/10.1111/1468-2427.12386.

González, R. (2015). El moviment per l'okupació i el moviment per l'habitatge: Semblances, diferències i confluències en temps de crisi. *Recerca, Revista de Pensament i Anàlisi, 17*, 85–106.

González, R. (2018). *Movimientos Sociales y Políticas Públicas: Los impactos de los centros sociales okupados en Cataluña y Madrid (1984–2014)*. Pachuca de Soto: Universidad Autónoma del Estado de Hidalgo.

Ibarra, P., Gomà, R., González, R., & Martí, S. (2002). Movimientos sociales, políticas públicas y democracia radical. Algunas cuestiones introductorias. In P. Ibarra, R. Gomà, & S. Martí (Eds.), *Creadores de democracia radical*.

Movimientos sociales y redes de políticas públicas (pp. 9–22). Barcelona: Ed. Icària.

Instituto Nacional de Estadística. (2011). *Censo de Población y Viviendas 2011*, http://www.ine.es/censos2011_datos/cen11_datos_inicio.htm. Date Accessed 14 November 2015.

Katz, S., & Mayer, M. (1983). Gimme Shelter: Self-Help Housing Struggles Within and Against the State in New York City and West Berlin. *International Journal of Urban and Regional Research, 9*(1), 15–45.

Kitschelt, H. (1996). Political Opportunity Structures and Political Protest: Anti-Nuclear Movements in Four Democracies. *British Journal of Political Science, 16*, 55–85.

Máiz, R. (1996). Nación de Breogán: Oportunidades políticas y estrategias enmarcadoras en el movimiento nacionalista gallego (1886–1986). *Revista de Estudios Políticos, 92*, 33–75.

Manrique, P. (2010, 8–21 July). Medidas penales contra las okupaciones. *Diagonal*, 22–23.

Martínez, M. (2002). *Okupaciones de viviendas y centros sociales. Autogestión, contracultura y conflictos urbanos*. Barcelona: Virus.

Martínez, M. (2007). El movimiento de okupaciones: Contracultura urbana y dinámicas alter-globalización. In R. Prieto (Ed.), *Revista de Estudios de Juventud, 76*, 225–243.

Martínez, M. (2016). Between Autonomy and Hybridity: Urban Struggles Within the 15m Movement in Madrid. In M. Mayer, C. Thörn, & H. Thörn (Eds.), *Urban Uprisings: Challenging the Neoliberal City in Europe* (pp. 253–281). London: Palgrave Macmillan-Springer.

Martínez, M. (2018a). The Politics of Squatting, Time Frames and Spatial Contexts. In M. Martínez (Ed.), *The Urban Politics of Squatters' Movements* (pp. 1–22). New York: Palgrave Macmillan.

Martínez, M. (2018b). Socio-Spatial Structures and Protest Cycles of Squatted Social Centres in Madrid. In M. Martínez (Ed.), *The Urban Politics of Squatters' Movements* (pp. 25–49). New York: Palgrave Macmillan.

Martínez, M., & García, A. (2013). Movimiento 15M, espacio público y luchas pro-vivienda. *Zainak, 36*, 87–105.

Martínez, M., & García, A. (2015). The Occupation of Squares and the Squatting of Buildings: Lessons from the Convergence of Two Social Movements. *ACME: An International E-Journal for Critical Geographies*, http://acme-journal.org/index.php/acme/article/view/1145/1107.

McAdam, D. (1998). Orígenes conceptuales, problemas actuales y direcciones futuras. In P. Ibarra & B. Tejerina (Eds.), *Los movimientos sociales. Transformaciones políticas y cambio cultural*. Madrid: Trotta.

Moreno, A. (coord.) (2012). *La transición de los jóvenes a la vida adulta. Crisis económica y emancipación tardía*. Barcelona: Obra Social La Caixa.

PAH (Plataforma de Afectados por la Hipoteca). (2018). http://afectadosporlahipoteca.com/.

Pastor, J. (2002). *¿Qué son los movimientos antiglobalización?*. Madrid: RBA Libros.

Peña-López, I. (2013). Casual Politics: From Slacktivism to Emergent Movements and Pattern Recognition. In J. Balcells, et al. (Eds.), *Bid Data: Chalenges and Opportunities* (pp. 339–359). Barcelona: Universitat Oberta de Catalunya. http://ictlogy.net/articles/20130626_ismael_pena-lopez_-_casual_politics_slacktivism_emergent_movements_pattern_recognition.pdf.

Pruijt, H. (2003). Is the Institutionalization of Urban Movements Inevitable? A Comparison of the Opportunities for Sustained Squatting in New Cork City and Amsterdam. *International Journal of Urban and Regional Research, 27*, 133–157.

Pruijt, H. (2004). Okupar en Europa. In R. Adell & M. Martínez (Eds.), *¿Dónde están las llaves? El movimiento okupa: Prácticas y contextos sociales* (pp. 35–60). Madrid: Los libros de la Catarata.

Romero, J. (2010). Construcción residencial y gobierno del territorio en España. De la burbuja especulativa a la recesión. Causas y Consecuencias. *Cuadernos Geográficos, 47*, 17–46.

Rootes, C. A. (1999). Political Opportunity Structures: Promise, Problems and Prospects. *La Lettre de la maison Française d'Oxford, 10*, 75–97.

Tarrow, S. (1986). Comparing Social Movement Participation in Western Europe and the United States: Problems, Uses, and a Proposal of Synthesis. *International Journal of Mass Emergencies and Disasters, 4*(2), 145–173.

Tarrow, S. (1997). *El poder en movimiento. Los movimientos sociales, la acción colectiva y la polític*. Madrid: Alianza Universidad.

Tarrow, S. (2010). The Strategy of Paired Comparison: Toward a Theory of Practice. *Comparative Political Studies, 43*(2), 230–259.

Tejerina, B. (1998). Los movimientos sociales y la acción colectiva. De la producción simbólica al cambio de valores. In P. Ibarra & B. Tejerina (Eds.), *Los movimientos sociales, transformaciones políticas y cambio cultural*. Madrid: Trotta.

CHAPTER 9

Squatted Social Centres Activists and 'Locally Unwanted Land Use' Movements in Italy: A Comparative Analysis Between Two Case Studies

Gianni Piazza and Federica Frazzetta

Introduction: Social Centres and LULU Mobilisations as Urban Social Movements

The importance of the conflicts and social movements in the cities is came back very timely, especially after the waves of protests in recent years since 2011 in the urban areas of the North and South of the world. Even scholars have focused new attention on the urban social movements, after the major works of Castells (1983) and Pickvance (2003), until recently to define them as 'conflict-oriented networks of informal relationships between individual and collective actors, based on collective identities, shared beliefs, and solidarity, which mobilise around urban issues through the frequent use of various forms of protest' (Andretta et al. 2015, pp. 202–203). Urban movements represent different ways to

G. Piazza (✉) · F. Frazzetta
University of Catania, Catania, Italy
e-mail: giannipiazza@tiscali.it

N. M. Yip et al. (eds.),
Contested Cities and Urban Activism, The Contemporary City,
https://doi.org/10.1007/978-981-13-1730-9_9

claim the 'Right to the City' that 'is far more than the individual liberty to access urban resources: it is the right to change ourselves by changing the city. It is, moreover, a common rather than an individual right since this transformation inevitably depends upon the exercise of a collective power to reshape the process of urbanisation' (Harvey 2008, p. 2).

Even in Italy, the long tradition of urban conflicts and movements has been renewed since the nineties thanks also to the decentralisation of the political system and the greater legitimacy given to the local administrations with the introduction of the direct election of the mayor. In fact, according to previous research (della Porta 2004), a decentralised political system may favour the emergence of urban movements. During the 1990s and first years of the new millennium, many urban conflicts emerged in which protesters also had the local administrators as allies against the central power, because the latter took decisions over urban territories without consulting local governments and citizens. Moreover, what previous research had always highlighted in the Italian case is that issues around environmental claims could have favoured alliances with left-wing parties, while issues around security claims could have had alliances with right-wing parties (ibid.). However, things have changed over the last ten–twelve years, and the difference between centre-right and centre-left local governments is no more so clear. Indeed, sometimes and often the centre-left administrations are the more closed to urban movements, especially to the more radical ones.[1]

In this paper, two types of urban movements largely diffused in Italy are taken into account, despite their differences: the first type is represented by the Squatted Social Centres (SCs), often described as more radical in terms of action repertoire and collective identity; the second type is related to LULU—Locally Unwanted Land Use conflicts in which citizens struggle for a 'better quality of life' (Andretta et al. 2015).

The squatting of SCs is a long-lasting phenomenon all over Europe: from the seventies to now, several hundred occupations have taken place, above all in southern European countries like Italy (Mudu 2004, 2012), Spain (Martìnez 2013), and Greece, but also in Germany, Holland, Great Britain, and Denmark (Piazza 2012; SQEK 2013;

[1]Although both authors share the contents of this chapter, Gianni Piazza wrote the paragraphs one–three and Federica Frazzetta the sections two–four, while they wrote together the concluding remarks.

SQEK, Cattaneo and Martinez 2014; Piazza and Genovese 2016). The SCs are usually empty and abandoned large buildings (former factories, cinemas, theaters, schools, and so on) which are occupied by left-wing radical and antagonist activists in order to self-organise and self-manage political, social, countercultural activities, and practicing participatory and non-hierarchical modes of political and social relationships (Piazza 2012). SCs cannot be considered only physical spaces, but also heterogeneous groups of squatters and activists that represent collective actors belonging to the radical left, that in Italy is dubbed 'antagonistic' left[2] (della Porta and Piazza 2008; Piazza 2012). Even if they are spatially localised in the city centres or in the peripheral and working-class districts, their reach of action is often not only local, but also regional, national, and global. In fact, claiming their political dimension, 'social center activists and squatters are thus often engaged in broader protest campaigns and social movements, fighting against precariousness, urban speculation, racism, neo-fascism, state repression, militarisation, war, LULU, private-oriented education/university reforms' (Martìnez 2013, p. 11).

Moreover, when the squatted SCs develop cooperative relations with other squatting and social movement organisations, they play a key political role inside the wider squatters' movement: they serve as an essential socio-spatial infrastructure for the coordination and public expression of the squatting as an autonomous urban movement (Martìnez 2013). Lastly, due to the radicalism of their claims (often not compatible with the logic of market and the institutional political system) and forms of action (often illegal and sometimes—but not always—confrontational with the police), the SCs are often labelled as violent by the mass media

[2] In Italy, the 'antagonistic left' is distinguished from the 'radical left'. On the one side, the 'antagonists' are those groups and activists belonging to these different political-ideological networks: autonomists, post-autonomists (ex *White Overalls-Disobedients*), Marxist-Leninists, anarchists, and so on, in which the Social Centers are divided, making the movement heterogeneous and often split (Mudu 2012; Piazza 2013). On the other side, the 'radical left' is composed by the political parties placed on the extreme left of the political-institutional spectrum, such as *SeL-Sinistra Italiana* (*Sinistra e Libertà*-Left and Freedom-Italian Left), now merged into the new political group *Liberi e Uguali* (Free and Equals); and those currently without representation in Parliament: PRC (*Partito della Rifondazione Comunista*-Refoundation Communist Party), within *Potere al Popolo!* (Power to the People!) PdCI (*Partito dei Comunisti Italiani*-Party of Italian Communists), and others (Piazza 2011).

and the politicians. They are indeed subjected to forced evictions and their activists can suffer the repression by the authorities (with complaints, arrests, and convictions). As we said previously, if some centre-left local governments have been friendly with urban movements, including the SCs during the nineties, nowadays it is very difficult to find a city administration (no matter if centre-left or centre-right-wing) open to SCs, with some exceptions.[3]

The other type of urban movement here considered is defined also as 'territorial' movement, because its reach of action can go beyond the urban areas, even if strictly interconnected with them. Just in the last decades, new territorial mobilisations have indeed emerged in Italy, but also in other countries. They are promoted by local communities and citizens' committees (della Porta 2004) that are giving voice to their claims through protest, opposing to unwanted urban and territorial transformations. What is at the stake is the use of the land, the territory as expression of a social conflict on the relative 'use value', or 'exchange value'. On the one side, the local political and economic elites consider the territory and the city as a mean to make profit or get consensus (exchange value); on the other side, groups of citizens and associations that deem the territory a use value for social and environmental purposes (della Porta and Piazza 2008).

These types of mobilisations are usually labelled by the media, politicians, and part of scientific literature as affected by 'NIMBY syndrome' (Not In My Back Yard), associated with a conservative behaviour and egotistical resistance to social change. They are interpreted as the refusal to pay the necessary costs (in terms of pollution, security, etc.) to attain public goods (Bobbio 1999) by few inhabitants who do not want public works in their territory, but would be indifferent if these infrastructures were made somewhere else. Other scholars prefer to define these phenomena in neutral terms, using the acronym LULU to indicate conflicts related to Locally Unwanted Land Use (della Porta and Piazza 2008; Andretta et al. 2015), that only registers the opposition to a certain use of territory not accepted by local population, without assessing a priori motivations and interpretative schemas. Still others scholars point

[3]For example, the city of Naples in the last years has been open to urban movements, including the SCs, but it is ruled since 2011 by a radical leftist government led by the mayor De Magistris, quite different from center-left administrations.

out how residents accused of 'nymbism', respond by building a Not On the Planet Earth (NOPE) discourse, that is affirming not to want disputed works 'either in their own, nor in any other backyard' of the Earth, because they are considered damaging for common good (Trom 1999). In fact, many of these conflicts are only seemingly parochial/localistic and/or environmentalist: large infrastructures, polluting plants, bases, and military installations are considered by the protesting local population, not only harmful to the environment and the public health, but also politically, socially, and economically. Moreover, many LULU conflicts become 'trans-territorial' movements: the protest actors are not only the local inhabitants, and they intertwine themselves with other extra-local players, building networks that go beyond territorial dimension, and showing propositional capacity, and not just reactive. Also, their political dimension 'goes beyond local boundaries because they involve institutional, political and social organisations that act on a regional, national and in some cases international level' (Piazza 2011, p. 329).

The collective actors involved in the LULU movements are socially and politically heterogeneous, they are linked by multiple memberships, and 'networked' in the course of action, thus producing further waves of mobilisation. Such networks involve diverse social groups, ideological viewpoints, and different generations, involving and re-defining the local community while remaining open to external actors. The types of collective actors that form into the networks, bringing their respective resources and skills while generating even some tensions, are very similar in the diverse LULU conflicts, although their relative weight and modes of interaction are rather different: citizens' committees, environmental associations, grassroots unions, and squatted SCs (della Porta and Piazza 2008; della Porta et al. 2013). Some LULU movements may have also local administrators (mayors, city councillors, and so on) as allies, in order to contest central government decisions taken without any local consultation. The distinction between centre-left and centre-right parties made by della Porta (2004) can be still significant if we look at the local level. However, if we look at the central state level, such differentiation among centre-left and centre-right governments is not always and explicative factor, and may happen that, for instance, national centre-left governments promote and sustain local project also against the opinion of environmentalist groups and the willing of local population (Piazza 2011).

Social Centres and LULU Movements in Italy

The participation of SCs in wider campaigns and social movements, like the territorial ones, comes either from their historical characteristics (such as the strong rootedness in the territory and in the urban fabric, and the close attention to the construction of alternative cultures), or from an already started path of political involvement in the territories—first for the area of the North East SCs, but then also for the autonomists and the anarchists (Panizza 2008; Piazza 2012). For what concerns the participation of SCs in LULU movements in Italy, the first evidences of such presence are dated after the Genoa G8 in 2001. Before that moment, SCs have been engaged mostly in city or neighborhood conflicts as central actors. The experiences of collaboration between SCs—or at least some of them—and other social movement organisations (including formal associations, parties, and unions) have multiplied in the course of the mobilisations of the Global Justice Movement (della Porta et al. 2006) and more recently in the mobilisations in the education systems and in the anti-austerity movements. After Genoa G8, the presence of the SCs activists[4] can be considered a regular feature in all or most of the LULU mobilisations emerged in recent years, where they can be deemed central actors together with other players (della Porta and Piazza 2008, 2016; Piazza 2011).

The SCs militants 'were always perceived by other Lulu protesters as allies, even if "troublesome" because of their stigmatisation by the media as extremist and violent. Above all, there have often been collaborative relationships between social centres and citizens' committees based on trust and mutual acknowledgement, thanks to continuous participation in the protest actions, even the riskiest ones, notwithstanding early distrust, tensions, and the ideological and cultural differences between them and both the unpoliticised citizens and the more moderate components' (Piazza 2011, p. 338). Their strong commitment in the mobilisation, but also in the consensual decision making processes, their experiences in the use of some forms of direct action, their reliability in accordance with the decisions taken, their availability to 'cross-fertilize' themselves encountering with others, have been recognised by their allies and the ordinary citizens, contributing to create bonds of mutual trust.

[4] Aware that the terms militants and activists are often used with different meaning, in this context we prefer to use them both as synonyms when referred to Social Centres.

Moreover, previous research had shown that the squatted SCs activists have brought to the LULU movements 'generational resources (young, mainly students), political-organisational resources (through the diffusion of No Tav and No Bridge[5] themes outside the local dimension) and repertoires of action (countercultural activities, creative demonstrations, direct actions) developed during the earlier occupation of autonomous spaces' (della Porta and Piazza 2008, p. 109). Here we mean as generational resources, those made by young generations that, to a greater extent than the older ones, have mobilised, have been attracted by the SCs and have preferred direct and disruptive forms of action in LULU movements. Furthermore, the antagonist left, including SCs, contributed to a 'scale shift' in the discourse of protest and especially to the creation of a network among the different mobilisations, transforming local conflicts into 'trans-territorial' (Piazza 2011). Indeed, while there are many territorial conflicts of different length and intensity, only some shift scale upward. This means that these conflicts expand in scale from the territory where the unwanted 'public bad' is located, toward other territories. The reach of action of the mobilisations and the political-institutional targets extend from the local to the regional, national, and—sometimes—transnational level (della Porta and Piazza 2008, 2016). In these cases, the SCs activists have contributed greatly to this scale shift favouring not only the link among the mobilizations coming from different territories, but also the connection among various issues and claims: defence of environment and health with struggles against war, for the social rights, for another development model, and so on (delta Porta and Piazza 2008, 2016). Nevertheless, the presence of SCs brought also elements of internal tension within the movements, most of all around the broader strategies to be adopted and the specific repertoires of action to be used (della Porta and Piazza 2008).

In the following pages, based on previous research (della Porta and Piazza 2008; Piazza 2011; della Porta et al. 2013) and current fieldwork—participant observation and semi-structured interviews—we focus on the role played by SCs activists within two LULU movements in Italy: those against the building of the high-speed railway in Val di Susa (No TAV) and the construction of a US Navy ground station of

[5]The No Bridge movement was aimed to prevent the construction of a bridge on the Messina Straits between Sicily and Calabria. The outcome of the protest was successful because the bridge was not built and the project stopped, at least until now.

satellite communications in Sicily (No MUOS).[6] We have chosen the No TAV and No MUOS because they can be considered as two of the most relevant LULU movements in Italy, because they became contentious issues for central and regional governments, for the importance they reached in the public opinion, the media attention they have had and the amount of publications and academic studies made on them (on No Tav, see: Caruso 2010; della Porta and Piazza 2008, 2016; Fedi and Mannarini 2008; on No Muos, see: Di Bella 2015; Mazzeo 2013). Moreover, even if they are both LULU movements, they are quite different experiences. In fact, the SCs taken into account are from three important urban centres (Turin, Catania, and Palermo), but they relate to two extra-urban areas (a mountain valley, the *Val di Susa* for the No TAV, and an agricultural area around a small town, Niscemi, for the No MUOS) with different socio-economic contexts. On the one hand, there is an industrialised area, with a developed social capital in the northern Italy; on the other, an economically underdeveloped area and with a scarce amount of social capital in the southern Italy (della Porta and Piazza 2008).

In a comparative perspective, we find interesting to investigate the same aspects in such different context in order to see similarities and differences. The interactions of the SCs militants with other movement groups and activists have been analysed, highlighting both the internal tensions and conflicting and cooperative relations. We have confirmed their ability to attract youth participation and to favour the cross-issues and cross-territorial scale shift, but we have also noted their ability to maintain the unity of the movement, notwithstanding the large difference with other groups. Moreover, the positive and negative feedbacks of their involvement in the LULU movements have been pointed out,

[6] In the case of the No Tav movement, our analysis is based first on four semi-structured interviews; then, even if our participant observations has not been continuous in time, we have taken part in the most important public assemblies and demonstrations in Val Susa. For what concerns the No Muos case, other four semi-structured interviews have been carried out, while the participant observation lasted from February 2013 to August 2014, during which marches, assemblies and meetings have been observed; moreover, we have participated in the main demonstrations from 2012 onwards. Lastly, all the interviews has been carried out, between January 2014 and September 2016, to the activists considered as key-informants for the SCs belonging to the two movements. Despite the key-informants interviewed are all males, the gender has not been a determinant variable in the choice of the activists to interview.

as well as the differences in terms of modalities of cross-fertilisation with other movement members and of geographical characteristics affecting the level of participation. Lastly, we have noted as a novelty, the ability of SCs activists to involve their 'urban constituency' in extra-urban mobilisations, in addition to that of cooperating with other collective actors, which usually they are in conflict and politically far from each other.

The Social Centres Activists in the No TAV Movement

The protest campaign against the construction of a 57km tunnel, as part of the TAV (*Treno Alta Velocità*—High Speed Rail Line) in Val di Susa, Piedmont, close to the border with France, originated in the 1990s (della Porta and Piazza 2008), but became more visible in the early 2000s, when the second Berlusconi government accelerated the policy-making process. The No TAV campaign, which started as a reaction to the risk of damage to the environment and health of citizens (due to soil erosion and asbestos within the mountain to be excavated), increasingly became proactive and constructive. The protest networks (made by environmentalists, local governments, citizens' committees, SCs, and grassroots unions) developed both specific alternative proposals and 'another model of possible development', based on 'de-growth' (Latouche 2007). Moreover, they claimed the right of the local population to decide the future of its own territory, demanding a different form of democracy, more participative and deliberative (della Porta and Piazza 2008). The development of such rich protest network has been possible also because the proximity of the metropolitan area of Turin with the Valley. In fact, Turin is not much far from some of the best-known villages of the Valley: less than 50 km from Bussoleno, about 50 km from Susa, about 60 km from Chiomonte. Moreover, the metropolitan area and many little villages of the Valley are well connected also by the public transport, such as trains and busses. This probably had helped the participation in the mobilisation of many organisations from the city, such as the SCs. The No TAV is the longest and best-known territorial movement in Italy, with the highest level of mass participation ever achieved in the country, and with the peaks of conflict and clashes with the police reached between 2011 and 2013, followed by complaints, arrests, and convictions for the No TAV activists. Currently, the mobilisation goes on, notwithstanding the state repression and the will of the Italian and French governments—at least until now—to continue and complete the building works (but they are far behind schedule and only the secondary tunnel is under construction). The new national government led by Conte in 2018 is split between the 5 Star Movement that is contrary to the TAV and the League which is instead favourable.

During the nineties, the SC Askatasuna[7] decided to join the No TAV movement. What interested the militants of the SC was the diversified composition of the movement, not exclusively formed by militants or by a specific social sector, like young students. The chance to live and build an intergenerational and transversal movement pushed the Askatasuna activists to join the No TAV struggle:

> We decided to invest in that field of struggle because we saw in that the possibility to break those mechanisms according to witch territorial movements where the expression of strongly ideological organised groups, also numerically limited, or that the large social movements belonged almost in an exclusive way to young people, with students struggles [...]. We have seen a fertile field, the possibility to say that territorial movements have not to be seen as a special corner for militants and also that mass social movements do not belong exclusively to some special sector of the society. [Int. No TAV A.]

Thanks to some militants and supporters coming from or still living in some villages of the Valley, the Askatasuna activists founded a grassroots committee, the Popular Struggle Committee (*Comitato di Lotta Popolare*) in Bussoleno. Through the committee, the SC militants were able to relate at the same level as the other No TAV committees and then to be recognised from the outset as an integral part of the movement:

> We founded with other comrades of Bussoleno, the Popular Struggle Committee, the first among the village committees, and unlike the others, it has always had a very political connotation, founded by people who already were used to politics. [...] At the beginning, it was mainly a mimicry approach, as we were used to said, in order to avoid to be [identified as] the social centres who come here ... to avoid to be too political, but try to blend into this strange subject, at that time still in formation, and

[7] Beyond the Askatasuna, the main SC in Turin, others were active—as *Gabrio, Asilo*, and *Murazzi*—but their activists were usual to participate in the No Tav movement not as representatives of their SCs. This is why we chose to take into account only the SC Askatasuna in our analysis.

> understand the features and the political potential for radicalism, without particular hurry, but trying to privilege and develop the behaviours and the attitudes that we saw in filigree but possible, without burning the stages, but at the same time with the clarity. I think it has been always our clear goal, I would say unpretentious and with humility from our point of view, to give priority to these attitudes here, these trends here and figure out when the time was right to try. [Int. No Tav G.2]

For what we had the opportunity to observe during the demonstrations and campsites, the SCs activists in Val di Susa have favoured the cross-issues scale shift. Indeed, they have connected and intertwined the defence of environment and citizens' health—the original issues—with others, as the struggles against the development model based on large infrastructures, and the 'corrupt-mafia system' that would have produced them (see also Piazza and Sorci 2017). Moreover, they have brought generational resources (mainly young students) to the No TAV movement, in the sense that many of the young people who participated in the movement were mobilised by the SC and attracted by its direct actions and countercultural practices. Then the activists of SCs have brought also and above all political-organisational resources, that is, the capacity to indicate the political direction towards which to move, the strategies of action to choose and to implement respect the policy of the institutional counterparty, as explained by one militant:

> I think that our contribution was of political knowledge, of organisational knowledge, of identifying the moments that counted and the points on which the conflict was outlined and to push on those points. I think we were able, with all the difficulties and mistakes of the case, to put before the eyes of the movement, or rather the most active part, what was the point on which there was the clash, the confrontation (with the opponent), the point on which one should not give in and one had to fight. This was our real contribution [Int. No Tav G.]
>
> (SC militants) had an important role in trying to protect the struggle from the countermeasures used by our enemies, first of all repression that we were used to know directly [...]. So having a certain amount of experience and trying to communicate it, we have participated without the aim to 'put the brand', an ideological imprinting on the struggle, but being available for the struggle. This has been the most important function that comrades had, most of all those from my organisation, the Askatasuna. [Int. No TAV A.]

The activists define themselves as very careful in dosing their forces and not 'forcing the hand', that is, not to propose political choices and strategies of action which would not have been understood and accepted by most of the other actors of the movement. Moreover, the Askatasuna militants are not the only ones in the radical sector of the movement, because some anarchist groups and activists participating in the No Tav campaign would have wanted to increase the level of clash with the police. In cases of conflict among the moderate and radical wings of the movement, the Askatasuna activists usually had a role of 'brake and control':

> That was our greatest contribution: we were able to measure out the times when the practice of rupture was put in the field without distorting the meaning of the (social and political) composition of the movement, without forcing it exceedingly. There was also a time when we had a brake role, on some components of the movement that came from outside and they wanted to do (force the actions) right away ... we tried to 'manage' these moments and I think we succeeded. [Int. No Tav G.]

Aware of the role played, the SCs activists used their political-organisational resources and abilities also to coordinate and maintain connected the different components of the movement, from the most radical to the most moderate, the institutional actors like the mayors of the Valley. They were conscious that their presence was useful in terms of political legitimation and there was not the risk they would affect the choice of the movement nor to institutionalise it:

> There are people who at certain times and contexts have an enormous weight, can determine things and, when it happens, may be unrecoverable and cause rifts. Instead, we have always worked for the unity of the movement, from mayors to the most radical forms. In our analysis, it was essential that there were mayors who continued to say 'no' at the institutional negotiation tables. Fundamental in Susa Valley were the No TAV electoral lists or No TAV activists in the local administrations. We do not elect anyone-the breaking with the White Overalls occurred right on the '*entrismo*' policy[8]-but we have a pragmatic look on the different situations, because

[8]The White Overalls were another political network of the Social Centers (see endnote 2)—above all in the North East Italy—coming from the *Autonomia*—as the Antagonist/Autonomist network of the SC *Askatasuna*—with which a political split took place in the

> it was an effective strength on the ground that the No TAV local administrators[9] had behind the movement. Those institutional figures, the mayors, were not able to affect the choices of the movement, but it was more the movement, which conditioned the mayors, from a certain point on. [Int. No Tav G.]

Such role is recognised also by other activists of the movement, that highlight how important is the contribution of Askatasuna in the No TAV struggle, as this interviewed says:

> SCs activists have been fundamental. [...] The first campsites, more than 16 years ago, have been strongly wanted by SCs, most of all by Askatasuna and then all the other SCs followed it. After all, here exists the Popular Struggle Committee, formed by Askatasuna's comrades and other people from the Valley, which have been a bonding agent among different subjectivities. Yes, I think that without the comrades of SCs this struggle would not be the same. [...] I mean, it is normal that SCs are a part of the movement. At the beginning maybe there was a bit of distrust, because of the way they are described by the bourgeois press. However, once they (SCs militants) started to deal with the Valley, they became habitants of the Valley to all intents. [Int. No Tav S.]

Thanks also to the contribution of the SC Askatasuna, the No Tav movement has had a cross-territorial scale shift, because it has become a point of reference for other social movements, not only the territorial ones, at national and international level. And also thanks to the SC activists, the movement has built bonds of solidarity with the activists of other territories and countries:

> today, this is an antagonist movement that feels close to all the other territorial movements, which feels close to all those things that move "against": against the status quo, against the government, against the institutional politics, without having an ideological "a priori" but having a method. [...]

1990s. The break was because WO chose the '*entrista*' policy, which was the strategic choice to 'enter' in the local institutions by participating to municipal election in alliance with the radical left-wing parties (Mudu 2012; Piazza 2013).

[9] For local administrators here we mean the mayors, the city councillors and cabinet members.

> almost everybody has come here: Palestinians, Kurdish, Greece, Basques. Most of all, the two summer campsites in 2011 and 2012 had an international participation by many militants: French, Dutch, Spanish people. In those two years, we had a European international dimension. Still today, antagonist people in Europe know what the No Tav movement is. [Int. No Tav G.2]

Through their participation in the mobilisation, the SCs militants increased their political strength and legitimation, thanks to their ability to enter in touch and cross-fertilise with other actors and above all with ordinary citizens, developing bond of mutual trust with them and contributing to train new activists and to politicise ordinary people:

> in Susa Valley a virtuous relationship has developed between the more antagonistic sectors (including SCs) and an average social composition of citizenry, often without political history, basically 'legalistic'. There was a process of encounter… the history has been more than ten years and that there have been real processes of mutual knowledge and trust (between antagonistic militants and ordinary people). What the movement gave us was quite a 'bath of reality': we grew up there as a political force, as recognisable as social and political weight on the city and on the national level, thanks to this movement; we had our history in the city, but it is clear that the movement has increased our strength. With the expression 'bath of reality' we mean that we have dealt with a social composition entirely different from us, that is, the elderly, parents, and so on. It has been very useful to deal with these differences and we managed to do what is much more difficult in the city… in Susa Valley it has formed a layer of militants of up to 50, 60, 70 years-old, that even now is in front of the gates in order to block the means of transport. We contributed and it is quite priceless, is the best and most interesting thing that there has been so far. [Int. No Tav G.]
>
> We grew up a lot and paradoxically one aspect is to learn also how to relate with the social fabric of the valley, but also with those political expressions like the local administrations and institutions. I say this because there has never been a monolithic approach; it has always bended according to the need, projects and strategies of the all movement. As I said before, the humility in learning and being available for such an important struggle led comrades not to approach in an ideological way to the territory, made by common people, comrades but also institutions. We approached in a non -– ideological way and we were able to decide, for instance, if a mayor could be the No TAV administrator and not the enemy to defeat, as we were used to in the city from which we came from. [Int. No Tav A.]

On the other hand, thanks to the No TAV movement, the SC Askatasuna increased its political weight and popularity at local and national level, as declared by an activist:

> We have had some benefits as experience and as capacity of individual and collective militants, and of course as a return of image and political rent on Turin and also on the national level. It (participation in No Tav movement) has undoubtedly allowed us to become a collective with its own strength, its recognition, being an actor more recognised, uncomfortable but also recognised in other areas of movement, because we had behind this story where we were inside, which we used, in addition to being used. Therefore, in that sense, we had a major feedback. [Int. No Tav G.2]

In addition, being such an active part of the No TAV movement let increase the level of police repression suffered by the Askatasuna militants:

> (Turin) became an extension of the Valley. The level of militarisation that there is in Turin is as such that, nowadays we can easily say that belonging to the Askatasuna or to students collectives means that you will surely deal with jail and precautionary measures, arrests and repression. This does not make us cry, it is just another responsibility. [Int. No TAV A.]

In fact, just to give some examples, in June 2011, few days before the eviction of the 'Free Republic of Maddalena',[10] the Turin Prosecutor issued 65 notices of investigation to the No Tav activists (among them, some from the Askatasuna). Simultaneously, police raided the Askatasuna by breaking down the door with a battering ram, in order to search the SC. More recently, eight activists of Popular Committee of Bussoleno have been undergone to home detention and four students of CUA-*Collettivo Universitario Autonomo* of Turin (Autonomous University Collective) were detained for a period and are still under

[10]On the 24 May 2011, thousands of No Tav activists occupied the ground in a mountain area named *Maddalena*, to prevent the start of the TAV building site. The occupation has been called "Free Republic of Maddalena" to claim the right of the people to decide on the use of land where they live. The occupation lasted until the 27 June, when a large deployment of police evicted the squatted land.

precautionary measures.[11] The police and judicial repression of the SCs activists has been consistent with the hostility and closure shown by most of political-institutional actors. Indeed, all the national and regional governments, both the centre-right and PD parliamentarians, the PD mayors of Turin have criminalised and stigmatised the No Tav militants coming from the SCs as violent and 'bad' protesters. Only the 5Stars politicians were and are opposed the Tav, but keeping the distance with the SCs militants. If this institutional enmity has certainly made the squatters' involvement in the mobilisation more difficult, it has not discouraged or prevented it anyway, thanks to their determination, to the support of local population and also to that—in some ways unexpected—of most No Tav mayors of Val di Susa.

The Social Centres Activists in the NO MUOS Movement

The movement against the MUOS (Mobile User Objective System) started in 2008–2009 in Niscemi (Sicily), with the main goal to stop the construction of a ground station of satellite communications inside the nearby US Navy base. If at the beginning the local residents had protested because they were worried about the health risks and environmental damage due to radio waves, very soon in the course of the mobilisation other protesters joined the town inhabitants, and their frames were extended beyond the concern for electromagnetic pollution: from NIMBY to NOPE activism. In fact, the struggle broadened in what the social movement scholars have called scale shift (Tarrow and McAdam 2005): activists from all over Sicily and Italy went to Niscemi to participate in national marches, 'struggling campsites', and direct actions. For them, the No MUOS became indeed a symbol of territorial resistance not only against unjust and nondemocratic decisions, but also against the war and the militarisation of the land. The protest network, formed by citizens' committees, SCs and antagonist groups, grassroots unions, and environmental associations, has been able to cross the territories from Niscemi to the regional, national, and transnational level.

[11] In December 2013, four No Tav activists were arrested on charges of terrorism (recently dropped) for having sabotaged a compressor. Even if the event has affected individuals that do not belong to the issue discussed herein, it is anyway important to report the event because of the gravity of the accusation.

Moreover the No MUOS were able to target different institutions and link with other similar LULU movements aimed at different uses of the territory (No Radar, No TAV, No dal Molin, No Bridge, and so on), and others movements as those for housing rights, for migrants rights, and so on (della Porta and Piazza 2014). The peaks of the conflict occurred between 2013 and 2014, when hundreds of activists blocked the road to prevent the entry of trucks into the base, being violently charged by the police, and thousands of demonstrators were able to invade and temporarily occupy twice the US base (129 activists have been charged for that). The Niscemi's station is now able to transmit, but thanks to the movement opposition, it happened with almost five years of delay in respect of US Navy projects. Despite this has led to a demobilisation phase, even recently the No MUOS continue to protest and demonstrate in Niscemi and in the cork forest around the US Navy base.

The No MUOS protest network gathered different groups and organisations from the main big cities of Sicily: Palermo, Catania, and Messina above all. Nevertheless, differently from the No TAV case, the Sicilian SCs activists coming from Catania and Palermo decided not to set up or to join any No MUOS committee, and not to enter in the Regional Coordination with the other local committees. They belong to the No MUOS movement as SCs, with their own identity recognised by all the other actors of the movement. In analysing the contributions that SCs brought to the No MUOS movement, such differentiation means that each player gave something different to the mobilisation:

> It is not possible to give a unique opinion on the role of these metropolitan groups (SCs). The point is that without their support, also in terms of numbers, it is not possible to think about No Muos struggle. They were able to bring a certain amount of social capital from the city, in terms of a huge amount of students that fill a coach or open a demonstration, in terms of the intelligent use of communication as *Officina Rebelde* does, in terms of radical forms of direct actions during demonstrations, when it is necessary (as it has been during the first and second campsite). It is an essential contribution. The problem is to find a point of political mediation [Int. No Muos F.]

Therefore, each SC has brought to the No MUOS movement different kind of contributions on the base of their identity, abilities, and possibilities. However, as it is evident from the abovementioned interview, with all their differences and specifications, the SCs have brought overall generational resources, as they attracted mostly young students and precarious workers, who were essential for the numerical strength of the demonstrations and their availability in the direct actions. Moreover, they have contributed with political-organisational resources (they were the main organisers of the permanent picket near the US base), and repertories of action based on countercultural activities, symbolic demonstrations, and above all disruptive actions. Such support and contribution changed during time, on the base of the movement's phase. In fact, the aim was not to impose or propose a political direction or a certain repertoire of action, regardless of the phase and movement's needs; the aim for the SCs activists was to support, with their own resources and political knowledge, the territorial movement, helping in the creation of a resistant community:

> I think that the method has always been the same, at least for what concerns us. We tried to valorise what was moving in the territory, because we think that in a territorial struggle what counts is the community that lives that territory. It is able not only to express needs but also to determine the balance of power towards the enemy, in this case US navy, the government and all of that carry on the MUOS project [...]. The role has been to push in the creation of an identity, of a community that from being atomised, divided in micro-structures, as we are used to know the society, become a resistant community, a community with a proper awareness and able to express its interests. [Int. No MUOS I.2]

This process happened not without internal tensions with the other more moderate sectors of the movement, but the SCs militants were able to preserve the unity of the movement in collaboration with the other activists. In fact, the internal tensions and conflicts occurred on strategies and forms of action between activists and group generationally and politically very different (the younger and more radical SCs militants vs. the older and more moderate pacifists and environmentalists). Nevertheless, these tensions were almost usually overcome through deliberative processes, cooperation, and division of labor (the ones more expert in direct action, the others more capable on legal and informative actions), above all in the peak of mobilisation. The different activists have used different means, according to their collective identity, without delegitimising those of the others. As you can notice by the surprised words of a senior pacifist interviewed:

After the 30 March 2013 demonstration, there was a contrast between the Regional Coordination of Committees and the Picket and antagonist activists, which led to the formation of two different decision-making arenas: the assembly of the Coordination (the grassroots committees) and the 'movement assembly' (the Coordination plus other antagonist activists and groups), in which internal conflicts were overcome by means of labour division and cooperation. In fact, consensual processes sometimes seemed to work well in delicate moments of direct action: on August 2013, there was (on that occasion) a political commitment of the most radical sectors, such as *Anomalia* ([Palermo SC militants, ed.]), with which we found the agreement and we managed to fix the points on the modalities and the management. And they took on the responsibility of the physical contact ([confrontation, ed.]) with the police, their faces uncovered and hands-free [...] I was shocked to see these activists, faces uncovered and hands raised, trying to pretend to break the blockade of the police, taking the truncheon blows in the face without blinking, while on the another side, other protesters were trying to get in, as it had been decided in order to divert attention. I thought it was a great sign of political maturity and quality in co-management of that action which was not taken for granted; ... I think that the police were neither prepared nor motivated to manage public order on that occasion and raise the level of repression [Int. No Muos A.]

The collaborative co-existence between the SCs activists, who pushed for direct action, and the other groups, who preferred the more moderate and legalist actions, is confirmed by the words of one SC Ex Karcere militant from Palermo:

One of the movement's richness is this: groups different from each other mainly promoted each of these different practices. However, in the moments of intense mobilisation these different practices have co-operated and co-existed in multiplying ways. For this, the effects of a single practice have had a more important weight when they were put together with those of the others. The different souls and sectors belonging to the No MUOS movement have their own peculiarities, their own interests and consolidated practices in which they are more expert and well-prepared; but the strong point of the movement was, in some moments, to be able to put together these different ways to understand and practice the opposition to the MUOS [Int. No Muos I.]

According to the activists, the participation into the No MUOS movement brought some positive aspects that directly involved SCs, such as the possibility to enter in the public debate, to enlarge their own network of relations but also to increase the field of action:

> There has been a positive feedback, as for all that struggles that have a clear aim and that can take those directions that can result fruitful for us, allow us to be present and talk in the public debate, to favour the conflict against the great economic powers that dominate our territory. It is good and allows us to have visibility, it gives us counter-power and allows us to intervene politically in Niscemi, but also in Palermo because of course if, to give an example, we have to fill a bus for a demonstration in Niscemi, we have the possibility to have contacts, build relationships, talk about political issues to the people we meet in the SCs and enlarge our field of action. [Int. No MUOS I.2]

Thanks to the contribution of all the movement sectors, the No MUOS movement intertwined relationships with other territorial movements. The SCs activists, in particular, contributed to the cross-territorial and cross-issues scale shift, also thanks to their preexisting links with the SCs militants involved in other LULU movements and to their tendency to generalise the local struggles and to connect different issues like the defence of the commons and the demand for democracy from below. An example was the common campaign shared by the No Tav, No Bridge, and No MUOS movements, during March 2013, against the 'exploitation of the territories'. This was highlighted in the following document, which launched the campaign:

> a month of mobilisations, in which we will have a unique voice and we will remember that the struggles against the bridge, the TAV, and the MUOS are naturally intertwined in a unique battle for the defence of commons. Our struggles have different specificities but there is a red thread that merges them in common goals and forms of actions that are empowered by a strong popular participation. Our struggles are strongly linked to those on labour and for the defence of rights, unfortunately united by the same repression, while it is still unheard the demand for democracy in which citizens can decide for their future.[12]

Even in the No MUOS case, the SC activists often have been subject to the repressive action by the authorities. Such as in 2014, when the

[12] "NO MUOS, NO PONTE, NO TAV Appeal": joint release of No Tav, No Bridge and No MUOS movements, 28 February 2013.

Gela Court on July 28 delivered 29 abode prohibitions to some No MUOS activists. In the document that explains the reason of this precautionary measure, the Court writes (referring to the demonstration that took place on 9 August 2013):

> Among the first rows of the demonstration, there were subjects ascribable to anarchist movements and Sicilian radical protesters, such as 'C.S.A. Ex Karcere' and 'C.S.A. Anomalia' from Palermo. 'Collettivo Aleph', 'Officina Rebelde', 'Collettivo Experia' and 'Teatro Coppola Occupato' from Catania and 'Teatro Pinelli' from Messina. Police already knew some of these activists, because they were present in previous and similar demonstrations.[13]

If the SCs militants have often been the target of the repressive police and judiciary actions as in Val di Susa, but with less intensity, they too have had few allies and many opponents in the institutional politics. From the U-turn of the Sicilian Governor Crocetta (PD), first opposed and then in favour of the MUOS, to the disinterest or hostility of many local politicians (excluding a shy support of the 5Stars), the SCs activists reacted strongly to these policies and attitudes, at least until before the most recent phase of demobilisation.

Concluding Remarks

In this contribution, we studied the field of relations among SCs activists and LULU movements in Italy, by analysing the case of No TAV and No MUOS movements. In investigating such field, we highlighted internal tensions, conflicting and cooperative relations between SCs activists and the other movement sectors, and the way in which they have been addressed. In the No TAV case, the SC Askatasuna is not the only radical actor of the movement, and therefore, it has often had a role of mediation in those conflicts between the moderate sectors and the most radical wings, like the anarchist groups. In the case of the No MUOS movement, the Palermo and Catania SCs have taken part in such conflicts with other movement sectors, also on the strategies and forms of action. Nevertheless, the activists all together managed to solve them, also thanks to the development of cooperative and trust relationships (at least in the increasing phases and the peaks of mobilisation).

[13] Order for the application of personal precautionary measures, Court of Gela, 16 July 2014.

Moreover, we confirmed previous research about the ability of the SCs activists to attract youth participation and favour the cross-issues and cross-territorial scale shift. Indeed, they were able to bring into the movement networks generational and relational resources, contributing to intertwine and make them closer to other social and territorial movements, at national and international level.

Even if the two cases studies taken into account in the analysis are quite different from certain points of view, some similarities emerged. Our work substantially confirms what was emerged in previous research, as regard the creation of mutual bonds of knowledge and trust, the contribution in training new activists, in bringing expertise in the countercultural activities, in the symbolic and direct forms of action, as well as bringing generational and political-organisational resources. In particular, in the No TAV movement, the role of co-leadership—with other players—of the SC Askatasuna of Turin emerged, as the activists contributed to indicate the 'political line' of the movement, the strategies of action to choose and to implement. Moreover, they have had the ability to maintain the unity of the movement coordinating and keeping connected the different sectors, from the most moderate to the most radicals, being aware that the unity of the movement was the necessary condition to hope for success. Also in the No MUOS mobilisation, the SCs militants acted to preserve the unity of the movement notwithstanding the differences with other groups. The internal tensions with the more moderate actors on strategies and forms of action were usually overcome through deliberative decision-making processes, cooperative relations and division of labor. The radical and antagonist militants were more involved in direct and disruptive actions, while the moderate eco-pacifists were more adapted to the informative and legal procedures, without anyone delegitimising the means used by the others. Therefore, the two analysed cases show that under certain conditions, like the land use at stake and the massive mobilisation attracting and involving unpoliticised citizen's participation, even the collective actors, which usually are in conflict and politically far from each other, can cooperate, struggle for a common goal, and find together shared strategies of action.

Moreover, in both cases the SCs activists recognise that from their participation in the territorial movements they had some positive and negative feedbacks and results. They evaluate as positive results the increased visibility and the new possibilities to enter in the public debate thanks to their participation in the territorial struggles. Moreover, they increased their field of influence in the cities where the SCs are based, and in the case of Askatasuna they also gained a stronger recognition at national level thanks to the No TAV movement. Lastly, in both cases, militants had the opportunity to deal with territories and social relationships quite diverse from what they were used to, and this increased their political experience. As negative results and feedbacks, if some openings in the political opportunity structure were valid for all the movement networks, in some cases the closure of such political opportunity windows involved mainly and directly the SCs activists. This is the case of some repressive operations, which involved mainly SCs activists with arrests, abode prohibitions, home detention, and other precautionary measures. Such focused repression measures had the aim to divide the two movements into 'good' and 'bad' protesters; in both No TAV and No MUOS cases, such attempts failed because most of the movements' participants refused such division and claimed each form of action (from the moderate to the most radical ones) as legitimate for the entire movement.

Over the similarities, there are some differences. First of all, Askatasuna and Sicilian SCs enter respectively in the No TAV and No MUOS movements in two different way: the former through the Popular Struggle Committee, the latter maintaining their SCs identities (notwithstanding Askatasuna from Turin and Ex Karcere from Palermo belong to the same ideological-political network). The 'cross-fertilisation in action' (della Porta and Piazza 2008, p. 9), which led to the creation of a 'hybrid' identity between SCs and the rest of the territorial movements, happened in different ways too. In the No TAV case, being a part of the movement as a grassroots committee, almost at the beginning of the campaign, posed the Askatasuna at the same level of the other local committees. In this way, not only the SC militants became a part of the diffused leadership of the movement, but they also shared with and acquired from the other sectors of the networks attitudes and forms of action. For example, a part of the Valley inhabitants became used to the confrontation with police and even to adopt radical forms of action, while the SC activists accepted the role of No TAV local administrators, by setting aside their conflictual attitudes toward institutions, which were grown in their political experience in Turin. In this way, the SC activists and the rest of the movement have built a 'hybrid' collective identity and a relation based on mutual trust. In the case of the

No MUOS movement, keeping their identity distinguished inside the movement did not compromise the creation of positive relations based on mutual trust with the other sectors and a common hybrid identity recognisable from the outside. While all participated in demonstrative actions, a sort of 'labour division' was created so that the SCs activists were able to manage mainly the radical forms of actions and protests, whereas the environmentalists and pacifist activists were in charge to follow—above all, but not only—the legal issues related to the MUOS and the counter-informative campaign. Such division did not lead to a reciprocal discredit; on the contrary, such division was at the base of the mutual trust. In such case, the hybrid identity was created with a level of intensity lower than the No TAV case, also because the SCs activists entered in the No MUOS movement network not at the beginning of the campaign but some years later, just before the peak period of mobilisation.

Then, what it has moreover to be underlined is the difference in terms of geographical characteristics affecting the level of participation. In the case of Susa Valley, its main villages—involved in TAV project—are quite near from Turin (less than 100 km); moreover, the Valley is well connected by public transports with Turin, as all the villages of the Valley among them. This, if apparently seems irrelevant, in reality may affect positively the level of engagement and favour the participation in the struggle. In the case of Niscemi, on the contrary, the site is quite far from Catania, Palermo, and Messina, and, above all, it is not well connected by public transport (railways transports do not exist and coaches are insufficient, if not quite absent). The long distances, with the lack of public transports, made more difficult for some SCs activists to be constantly present in Niscemi and even affected the level of their engagement and participation in the struggle.

Lastly, we would highlight what we believe quite interesting in this contribution, even if it would require further insights. What is quite significant and new, comparing previous research, is that the SCs, in relating with territorial movements, have been able to involve that part of society that they consider their urban 'reference' or 'constituency' in the extra-urban places and struggles. Urban citizens and activists, mainly students and precarious workers, have all been involved in the territorial struggle outside the urban setting. This is one of the main results of the operation of cross-fertilisation and cross-territorialisation among SCs activists and LULU movements. The other is the ability to cooperate and to mobilise together for a common goal with the activists of groups and organisations, with whom they usually have difficult relationships, of non-collaboration or even open conflict.

References

Andretta, M., Piazza, G., & Subirats A. (2015). Urban Dynamics and Social Movements. In D. della Porta & M. Diani (Eds.), *The Oxford Handbook of Social Movements* (pp. 200–215). Oxford: Oxford University Press.

Bobbio, L. (1999). Un processo equo per una localizzazione equa. In L. Bobbio & A. Zeppetella (Eds.), *Perché proprio qui? Grandi opere e opposizioni locali.* Milan: Franco Angeli.

Caruso, L. (2010). *Il territorio della politica. La nuova partecipazione di massa nei movimenti No Tav e No Dal Molin.* Milan: Franco Angeli.

Castells, M. (1983). *The City and the Grassroots: A Cross-Cultural Theory of Urban Social Movements.* Berkley: University of California Press.

della Porta, D. (Ed.). (2004). *Comitati di cittadini e democrazia urbana.* Soveria Mannelli: Rubettino.

della Porta, D., & Piazza, G. (2008). *Voices of the Valley, Voices of the Straits. How Protest Creates Communities.* Oxford and New York: Berghahn Books.

della Porta, D., & Piazza, G. (2014). *Scale Shifting in Territorial Conflicts: The No Muos Movement from the Local to the Global.* Paper Presented at the SISP Annual Conference, University of Perugia, 11–13 September.

della Porta, D., & Piazza, G. (2016). Il cambiamento di scala del Movimento No MUOS: oltre la protesta contro l'inquinamento elettromagnetico. *StrumentiRes. Rivista online della Fondazione RES, 2*(7), 1–28.

della Porta, D., Andretta, M., Mosca, L., & Reiter, H. (Eds.). (2006). *Globalization From Below: Transnational Activists and Protest Networks.* Minneapolis and London: University of Minnesota Press.

della Porta, D., Fabbri, M., & Piazza, G. (2013). Putting Protest in Place. Contested and Liberated Spaces in Three Campaigns. In W. Nicholls, B. Miller, & J. Beaumont (Eds.), *Spaces of Contention: Spatialities and Social Movements* (pp. 27–46). Farnham: Ashgate.

Di Bella, A. (2015). The Sicilian MUOS Ground Station Conflict: On US Geopolitics in the Mediterranean and Geographies of Resistance. *Geopolitics, 20*(2), 333–353.

Fedi, A., & Mannarini, T. (Eds.). (2008). *Oltre il Nimby. La dimensione psicosociale della protesta contro le opere sgradite.* Milan: Franco Angeli.

Harvey, D. (2008). The Right to the City. *New Left Review, 53,* 23–40.

Latouche, S. (2007). *Petit traité de la décroissance sereine.* Paris: Mille et Une Nuits.

Martìnez, M. (2013). The Squatters' Movement in Europe: A Durable Struggle for Social Autonomy in Urban Politics. *Antipode, 45*(4), 866–887.

Martìnez, M., & Cattaneo, C. (2014). Conclusions. In SQEK, C. Cattaneo, & M. Martìnez (Eds.), *The Squatters' Movement in Europe* (pp. 237–249). London: Pluto Press.

Mazzeo, A. (2013). *Il MUOStro di Niscemi. Per le guerre globali del XXI secolo.* Florence: Editpress.

Mudu, P. (2004). Resisting and Challenging Neoliberalism. The Development of Italian Social Centers. *Antipode, 36*(5), 917–941.

Mudu, P. (2012). I Centri Sociali italiani: verso tre decadi di occupazioni e spazi autogestiti. In G. Piazza (Ed.), *Il movimento delle occupazioni di* squat *e centri sociali in Europa*. Special Issue of *Partecipazione e Conflitto, 4*(1), 69–82.

Mudu, P. (2014). "Ogni Sfratto Sarà una Barricata": Squatting for Housing and Social Conflict in Rome. In SQEK, C. Cattaneo, & M. Martìnez (Eds.), *The Squatters' Movement in Europe* (pp. 136–163). London: Pluto Press.

Panizza, C. (2008). Grandi opere e protesta: sindrome di Nimby o riappropriazione della politica? Intervista a Donatella della Porta e Gianni Piazza. *Quaderno di Storia Contemporanea, XXXI*(44), 89–103.

Piazza, G. (2011). "Locally Unwanted Land Use" Movements: The Role of Left-Wing Parties and Groups in Trans-Territorial Conflicts in Italy. *Modern Italy, 16*(3), 329–344.

Piazza, G. (2012). Il movimento delle occupazioni di *squat* e centri sociali in Europa: Una Introduzione. In G. Piazza (Ed.), *Il movimento delle occupazioni di* squat *e centri sociali in Europa*. Special Issue of *Partecipazione e Conflitto, 4*(1), 4–18.

Piazza, G. (2013). How Do Activists Make Decisions Within Social Centres? A Comparative Study in an Italian City. In SQEK (Ed.), *Squatting in Europe: Radical Spaces, Urban Struggles* (pp. 89–111). New York: Autonomedia.

Piazza, G., & Genovese, V. (2016). Between Political Opportunities and Strategic Dilemmas: The Choice of "Double Track" by the Activists of an Occupied Social Centre in Italy. *Social Movement Studies*, Online Publication Date: 23 February 2016. https://doi.org/10.1080/14742837.2016.1144505.

Piazza, G., & Sorci, G. (2017). Do LULU Movements in Italy Fight Mafia and Corruption? Framing Processes and 'Anti-System' Struggles in the No Tav, No Bridge and No Muos Case Studies. *Partecipazione e Conflitto, 10*(3), 747–772.

Pickvance, C. (2003). Symposium on Urban Movements. *International Journal of Urban and Regional Research, 27*(1), 102–177.

SQEK. (2013). *Squatting in Europe. Radical Spaces, Urban Struggles.* New York: Minorcompositions/Autonomedia.

SQEK, Cattaneo, C., & Martìnez, M. (Eds.). (2014). *The Squatters' Movement in Europe. Commons and Autonomy as Alternative to Capitalism*. London: Pluto Press.

Tarrow, S., & McAdam, D. (2005). Scale Shift in Transnational Contention. In D. della Porta & S. Tarrow (Eds.), *Transnational Protest and Global Activism* (pp. 121–150). Lanham: Rowman & Littlefield.

Trom, D. (1999). De la réfutation de l'effet NIMBY considérée comme une pratique militante. Notes pour une approche pragmatique de l'activité revendicative. *Revue Francaise de sciences politique, 49*, 31–50.

Interviews

Int. No Tav A.: Interview with Andrea, Askatasuna Social Centre Militant, Venaus, 30 July 2016, Carried Out by Alessandro Rapisarda.

Int. No Tav G.: Interview with Gianluca, Askatasuna Social Centre Militant, Torino, 6 June 2015, Carried Out by Massimiliano Andretta.

Int. No Tav G.2: Interview with Gianluca, Askatasuna Social Centre Militant, Venaus, 24 July 2016, Carried Out by Alessandro Rapisarda.

Int. No Tav S.: Interview with Stefano, No Tav Activist, Venaus, 30 July 2016, Carried Out by Alessandro Rapisarda.

Int. No Muos A.: Interview with Antonio, Reporter and Antimilitarist Activist, Messina, 17 January 2014, Carried Out by Federica Frazzetta.

Int. No Muos F.: Interview with Federico, Officina Rebelde Catania Militant, Catania, 18 August 2015, Carried Out by Federica Frazzetta.

Int. No Muos I.: Interview with Ivan, Ex Karcere Social Centre Militant, Palermo, 14 January 2014, Carried Out by Federica Frazzetta.

Int. No Muos I.2: Interview with Ivan, Ex Karcere Social Centre Militant, Palermo, 24 September 2016, Carried Out by Alessandro Rapisarda.

PART IV

Urban Activism—Citizenship and Right to the City

CHAPTER 10

'We Are Quality Citizens of Bangkok Too': Urban Activism in Bangkok During the 2011 Floods

Danny Marks

Introduction

In the second half of 2011, Bangkok experienced its worst flooding in many decades. Overall in Thailand, the floods killed over 800 people, affected millions and cost the economy at least US$45 billion. Much of this devastation occurred in Bangkok and its environs. The 2011 floods also exposed vast inequalities in Bangkok in terms of those who were exposed and those who were protected. During the floods, some died, and some were injured and/or became sick, while others experienced financial and emotional distress. Still, others were barely flooded while some stayed dry. Reflecting Bangkok's socioeconomic and political inequalities, communities in the peri-urban fringes, particularly slum communities, were inundated the most and the longest, whereas the inner city of Bangkok, the home of the central business district and many of

D. Marks (✉)
Department of Asian and International Studies,
City University of Hong Kong, Kowloon Tong, Hong Kong
e-mail: danny.marks@cityu.edu.hk

N. M. Yip et al. (eds.),
Contested Cities and Urban Activism, The Contemporary City,
https://doi.org/10.1007/978-981-13-1730-9_10

the elite, and the city's industrial estates were protected by the national government (Marks 2015).

During the floods, many local communities felt a deep sense of injustice in response to the state's actions and challenged this protection of the elite and the industrial estates. They demanded more equal redistribution of floodwater and compensation if they are flooded. In 2011, 85 flood-related protested occurred in the Bangkok Metropolitan Region (BMR), most of them over the location and height of temporary sandbags walls, the operation of water gates and compensation schemes after the floods subsided.

By looking at protests against the state's responses to flooding, this chapter considers a rarely discussed form of urban activism in the Global South. Scholars have written about urban activism in the Global South which arose in response to forced evictions, dispossessions and demolitions of houses (Sheppard et al. 2015), incinerators which would be built in Chinese cities (Wong 2016), removal of vendors from the streets and plazas in Mexico City (Walker 2013) and the push for more democratisation and autonomy in Hong Kong (Ortmann 2015). However, very little has been written so far about urban activism during disasters and the effects of this activism on claims to the right to the city.

To address this gap, this chapter first summarises the existing literature on urban activism against disasters, the concepts of environmental justice (EJ) and the claim to the right to the city. It next discusses the state's inept and unequal response to the 2011 floods during the floods and the activism that arose in response to the state's actions. This chapter then describes two case studies of activist communities in Bangkok which heavily protested against the uneven shape of the flood during 2011. This chapter concludes by discussing similarities between the cases and how these findings contribute to the literature on urban activism. The findings are based on sixteen months of fieldwork conducted in Bangkok from 2014 to 2015 which included 100 interviews of residents whose houses were flooded.

Connecting the Right to City to Flood Protests

This chapter seeks to examine how protests against the state's governance of disasters, such as flooding, relate to the concept of the right to the city. It invokes the arguments laid down by urban political ecology (UPE) and EJ scholars. With a strong Marxist leaning, UPE developed from the work of Harvey (and Lefebvre). In his seminal work, Social Justice and the City, Harvey begins with the position that the city is a tangible, built environment but also a social product. Cities are built for the purpose of circulating capital, including human capital, commodities or finance. Using this Marxist framework, he argues that 'cities are founded upon the exploitation of the many by the few' (Harvey 1973, p. 314) and posits that the roots of urban inequality are the scarcity and high value of land in good locations. Urban political ecologists expand upon Harvey's theory of the city, perceiving landscapes and urban infrastructures of cities as hybrids and 'historical products of human-nature interaction' (Keil 2003, p. 724). Thinking of the city as a socio-spatial hybrid enables us to see how the 'social production of urban space unevenly spreads the vulnerability to hazards, exposure to risk and ecological breakdown' (Murray 2009, p. 171). Spaces of environmental degradation and high exposure to hazards as well as those of protection to hazard threats are unevenly distributed over the topography of the city.

Specifically, as Collins argues (2010) in the case of urban flooding, uneven vulnerabilities experienced by different individuals during floods are largely due to the state and market institutions protecting the lives and the interests of the elite while failing to protect marginalised groups or making them more vulnerable. Normally, the state, rather than the private sector, has undertaken investments in flood risk reduction, such as flood protection structures, designation of public floodways and land-use controls and therefore plays a key role in determining how vulnerable people are to floods. Hence, the state is a crucial arena of contestation over flood protection. In this contemporary landscape, the elites often have been able to use the state to accumulate social surpluses in the spaces where they live and work at the expense of other groups (Collins 2010). However, the structures of power governing floods and their effects are not static. Rather, by opening political space, floods can act as catalysts or tipping points shaping 'the future political trajectory towards an accelerated status quo or a critical juncture' (Pelling and Dill 2009, p. 29).

Environmental justice (EJ) scholars add that socioeconomic processes create unjust vulnerabilities to flooding. They argue that the 'distribution of environmental goods and harms' has a tendency to 'follow that of economic goods and harms, particularly within cities' (MacCallum et al. 2011, p. 1). Schlosberg (2007) theorises that there are three types of environmental injustices: distributive, procedural (whether different groups have equal access to decision-making regarding the environment) and lack of recognition (whether groups have been discriminated against due to their identity). EJ analyses should therefore compare the spatiality of environmental harms, such as sites of toxic hot spots in communities of colour or, in this case, sites of exposure to floods.

Given that EJ emphasises that victims of urban disasters have rights which should be realised, this next section discusses the agency of such victims in seeking to address these injustices. Some authors have analysed the linkages between activism and disasters. For example, Sokefeld (2012) describes public political action and mobilisation at the national level in response to a landslide in a village in Pakistan. Jain (2010) argues that damages caused by flooding and winter storms is a key motivating factor behind community protests in South Africa. Roth and Warner (2007) discuss protests against planned polder construction in The Netherlands. However, thus far, little has been written about urban activism during disasters and the how such activism is a claim to a right to the city. Therefore, my research (especially Chapter 8) helps to fill this gap.

It is first necessary to review what Lefebvre meant by the right to the city. He wrote, 'The right to the city cannot be considered a simple visiting right or a return to the traditional city. It can only be formulated as the right to urban life, in a transformed and renewed form' (1993, p. 435). It is the right of not only property owners, but of all people living in the city, in particular those who are currently without this right (Marcuse 2012). Harvey adds, 'The freedom of the city is, therefore, far more than a right of access to what already exists: it is a right to change it more after our hearts' desire' (2009, p. 45). Marcuse envisions that this is a moral right to a better system, a system in which the benefits of urban life can be fully obtained. He declares, 'it is a right to social justice' (2012, p. 34). However, the right to the city does not imply a clean and charming city, frequenting lovely parks and living there peacefully (Dikec 2009). Instead, as Lefebvre asserts, '[I]t does not abolish confrontations and struggles. On the contrary'! (1996, p. 195). The city should be a place full of encounters and an arena to redefine rights because this right to exercise a change in urban life can only be claimed through collective mobilisation and political struggle.

Dikec usefully argues that the concept of spatial justice should be considered when conceptualising the right to the city. This is because forms of injustice, such as domination and marginalisation, manifest themselves spatially, not only in the built environment, but also in less visible spaces of flows and networks. The right to the city is also a 'reconsideration of the spatial dynamics that make the city' (Dikec 2009, p. 83). Consequently, spatial dynamics of a city must be reconsidered and when a city's inhabitants seek to claim their right to the city, the spaces of injustice must be emancipated collectively. This emphasis on spatiality ties well to the ways in which political ecologists view disasters.

If the concept of the right to the city means, as Marcuse (2012) proclaims, a right to a better system in which the benefits of urban life can be fully obtained and to social justice, then the concept must encompass additional rights. Since the effects of disasters are the result of sociopolitical processes, the right to equal and just protection from or mitigation of disasters, such as floods, is part of this right. If some of a city's inhabitants suffer severe flooding or droughts, as urban political ecologists argue, whereas others do not, then all citizens' rights to the city are not being fully realised. Further, as Lefebvre (1993) and Harvey (2009) assert, if this right to the city must be collectively fought for, then collective action against those who deny rights to just and fair disaster risk mitigation is a form of claiming this right. Last, since, as Dikec (2009) argues, spatiality is a key concept of the right to the city, if the spatiality of vulnerability to a disaster is uneven and unjust, spatiality must be contested and the effects redistributed. Therefore, collective activism against the state's unjust responses to flooding should be seen as a claim of the right to the city. The rest of this chapter discusses how protests in Bangkok against the state during 2011 floods sought to reshape the projected spatiality of future floods and to make a claim to the right to the city.

The 2011 Floods: Causes and Response of the State and Local Communities

Before discussing the protests, it is necessary to review the 2011 floods and the sources of the protesters' aggravation. What made this flood unusual and 'unnatural' is that it was a slow onset event. Since the Central Plains is relatively flat with a gentle gradient, once water overflowed river banks in this area, the front of the floodwater advanced slowly southward, at only 2–3 km per day. This meant that significant time existed for proposals and counterproposals, actions and reactions, as well as deadlocks, to mitigate the impact of the anticipated flooding downstream. For the BMR itself, the national and local government agencies had perhaps a few weeks to prepare, but it could not be predicted how much water would hit the BMR and when it would arrive.

Moreover, Thailand lacked an accurate and well-coordinated flood data collecting and warning system: no central data system existed to gather and analyse flood information. Many agencies were involved in collecting water data but did not coordinate with each other (Marks and Lebel 2016).

Dykes, roads and temporary floodwalls created highly uneven exposure to the floods. Once the floodwater hit Bangkok, the Bangkok Metropolitan Administration (BMA) constructed huge sandbag barriers, closed water gates and diverted water to the west to protect the city's central districts. As a result, reflecting Bangkok's socioeconomic and political inequalities, those living in the peri-urban fringes were inundated the most and the longest, and their assets were heavily damaged. In contrast, the inner city of Bangkok, the home of the central business districts and many of the elite and the city's industrial estates remained dry and were protected by the national and Bangkok metropolitan government. This decision generated significant discontent among local residents in these areas, who had seen on the news or heard from others that the inner city was still dry, but their area had been flooded for weeks. Or they saw that houses located on the other side of the sandbag walls and water gates were hardly flooded whereas their homes were submerged in up to two metres of water.

In response, throughout October and November, these residents frequently expressed their anger through petitions and protests and attempted to destroy the sandbags or open water gates. For example, in the early November, after enduring protracted inundation, hundreds of residents of a housing estate in western Bangkok blocked a major road, insisting upon the removal of a sandbag barrier and only dispersed after the police agreed to remove the barrier (*Bangkok Post* 2011a). In mid-November, over 200 residents in Pathum Thani removed the sandbags by Khlong Hok Wa as well as demanded a nearby water gate to be opened by 20 cm (see section "Kalijawi and Warga Berdaya" in Chapter 7) (Krungthep Turakij 2011). Around the same time, almost a 1000 residents in Don Muang demolished a sandbag dyke after the government reneged on its promise to lower the level of the dyke (see section "Sociopolitical Context After 1998" in Chapter 7) (*Bangkok Post* 2011b). On 24 November, in the western area of Bang Khae, residents blocked a section of the Western Outer Ring Road, demanding that the government explain its unfair flood mitigation measures. They said that they had to endure chest-high putrid floodwater for over five weeks without any explanations or response from the government (*The Nation* 2011). On the same day in western Bangkok, around 100 residents from Thawi Watthana, Taling Chan and Bang Kae districts closed a section of Kanchanapisek road Nang and demanded that City Hall clarify its flood management plan. 'We are willing to ease the hardship of people in Nonthaburi, said Rueng Muangchum, a resident of one of these districts. We just want to know what the BMA's drainage plan is, how they will drain the water and how long the flood will stay' (Frederickson 2011).

Further examples abound. The last two protests described above suggest that an additional problem was that the government did not communicate clearly its flood management plans with communities. The head of a local NGO complained during an interview that the government neither articulated to these communities its method to drain the water in flooded areas, the duration these areas would be flooded and the location of where it would place sandbags. Nor did it give strong rationales for its mitigation measures. This lack of clear information frustrated residents because they did not know how much longer their areas would stay flooded and why their area had remained flooded while others remained dry.

During the floods, many of those who were flooded did not remain passive victims. Instead, they took a number of actions at the household level to reduce their own vulnerability, such as evacuating, moving their possessions higher or using boats to buy supplies and receive relief. Moreover, while communities in Thailand are social and administrative constructs and have conflicts and divisions within them (Elinoff 2013), most communities in the BMR became unified and worked together to contest the scalar configurations of the floods to reduce the level of water and shift the risk of the floods to other areas. Consequently, numerous flood-related protests occurred in the BMR, most of them over the location and height of temporary sandbags walls, the operation of water gates and compensation schemes after the floods. Overall, according to a former Flood Relief Operation Centre (FROC) official, 85 flood-related protests occurred in the BMR.[1] The next two sections are case studies of

[1]They were 67% of the total number of flood-related protests (126) in the country. According to this official, his staff members obtained this number by counting the number of protests mentioned in Thai-language newspapers. However, the criteria for what was considered a protest by newspapers are unclear. I never received a clear answer to this question during my interview.

communities in Bangkok who felt that the state's response to the floods was unfair and challenged the spatiality of the flooding.

Yucharoen Community: Challenging Inner City Dry and Outer City Flooded

Located on the northern edge of Bangkok, Don Muang is one of the districts which was the worst affected by the floods. Overall, 84 communities and 65 housing estates were flooded (*Bangkok Post* 2011d). Don Muang today has the fourth-highest population among districts in Bangkok: 166,210 people in 2011 according to the Bangkok Metropolitan Administration (2012). Yucharoen community, a lower-middle-income community is located in northern Don Muang (see Fig. 10.1). All of the residents live in Yucharoen Housing Estate, one of many estates built on farmland in the 1980s. The houses are two-story row houses each narrowly separated from each other. Most residents are not from Bangkok but recently moved to Bangkok and bought a row house in the community. Many work at the airport or nearby.

The residents felt a deep sense of injustice about placement of a wall of big sandbags by FROC, which was the national government's flood response centre. Comprising thousands of giant sand bags, the barrier ran from the area north of Don Muang airport to Khlong Sam Wa. Each 'big' bag weighed 2.5 tonnes (*Bangkok Post* 2011a). According to Tinnakorn, the Yucharoen community leader, FROC did not tell the communities residing outside the wall that it was going to place the bags there before it had. Nor did FROC tell them how long the bags would stay and FROC's rationale for placing the bags where it did.

Immediately, Yucharoen residents became incensed about the placement of this wall which blocked water from leaving their community while the inner city of Bangkok stayed dry. Their area was declared a disaster zone and, as Tinnakorn described, 'the old water stayed and the new water came.' As a result, the water became stale, dirty and full of mosquitos. Tinnakorn soon gathered a group of those who had not evacuated from Yucharoen to negotiate with FROC regarding the stale water, assistance (the wall made it difficult for the community to receive assistance) and ability to travel (since the wall also blocked them from travelling to most of Bangkok). However, FROC did not listen and agree to their demands, and so, they decided, as Tinnakorn declared,

Fig. 10.1 Map of Yucharoen in Don Muang (*Source* Map from Google Earth under creative commons licence)

'to fight FROC.' Yucharoen residents rallied residents of other affected communities in Don Muang living outside of the wall to join them to protest against the wall so that they could improve their negotiation position.

Led by Tinnakorn, around 500 people from an estimated 20 communities, including those in Phrom Samrit (see Fig. 10.1), throughout Don Muang protested thrice at the big bag site. They were one of the first groups in Bangkok to protest against the government's response to the floods. During the first protest, they rode on boats to the wall and then held a public hearing and a symbolic vote on whether or not to remove the sandbags. The result was a unanimous decision for removal. 'We are quality citizens of Bangkok too' was their repeated slogan, as they demanded that the government recognise not only those living in the inner city but also them as members of Bangkok. They also used the death of an elderly man who had drowned to suggest that it was not too safe to live in this area as long as water was being blocked by the wall. They then demanded that FROC respect their constitutional and human rights, including (1) the right of movement since the barrier blocked transportation along Viphawadi-Rangsit Road, (2) the right to their livelihood since the wall blocked from receiving food supplies, and (3) the right to have a public hearing about the barrier. After legally pressuring the government and threatening to close the tollway which runs above Viphawadi-Rangsit Road, they called the government to negotiate. Unbeknown to the government, a TV reporter arrived to capture the negotiations between Tinnakorn and FROC's negotiator, Karun Hosakul (a local Member of Parliament) on television. Being live on air, Karun felt pressured and after listening to their demands, he conferred with the head of FROC and subsequently promised that FROC would remove the bags within a couple of days. However, when that day came, the government did not send anybody to remove the bags, so the people felt that they had the legitimate right to break down the wall. And so, they did it themselves after the deadline had passed by using their knives to cut through the bags, so the water would melt the sand. The police did not stop them. Their protests gained the attention of the media: this story was covered by the Bangkok Post, Thai PBS, and other news agencies. Tinnakorn was mentioned and quoted in these stories.

The level of the water sank rapidly (approximately seven inches according to Tinnakorn) once the bags were removed, and the quality of the water in Yucharoen also improved because the older water flowed out through the opening in the barrier. The dismantling of the barrier was helpful because it enabled them to travel easier and improved access for government agencies to distribute flood relief items to them, including food and water.

The protesters were very proud that such a small organisation could directly negotiate with the government and the government listened to their demands. They were one of the first groups of protesters in BMR to, using Tinnakorn's language, 'take out the big bags.' Consequently, they believed that they had influenced protesters in other areas of Bangkok to use similar tactics and make similar demands. He remarked, 'People in those areas disobeyed the orders of the government by using the same constitutional argument we did. They used the same strategy of closing the road in Rama to request for more assistance or to pressure the government to open the water gate.' He also added that FROC 'gave more assistance and relief than normal' to this area because 'of our past demands.'

Twenty-five interviews conducted with Yucharoen residents reveal that the majority of the interviewees (around 78%) either joined or supported the protests. Of the 23 people who answered questions about the protests, eight attended, five did not. In addition, eight people who evacuated said they would have protested if they were there, and two others supported the protests but could not join. Those who joined the protests gave the following reasons for doing so:

- "Yes, I participated. I decided to protest because it was not fair. I went to show that I was not happy with the situation."
- "The big bags were not fair. I was on the boat waiting for the government. I was angry at the government. I can't accept it-I insulted government representatives."
- "I protested because we needed to help each other-I couldn't stand the smelly water anymore."
- "I protested because I was badly affected-there were mosquitos and the water was very bad quality. It was no not convenient for me to stay so I protested against the big bags. I protested against the Bangkok governor because he was the one who didn't release the water. The PM ordered to open the bags but the governor didn't know. I saw this from the TV news."

Most of those who protested or stayed in the community agreed with Tinnakorn that the protests were successful. They believed that the water level lowered and the water quality improved. For example, a retired soldier in the community proclaimed, 'As soon as the big bags were gone, the water level quickly dropped.' Some also expressed their appreciation towards their community leader for leading the protests. For example, one woman who had evacuated, stated, 'I think it's good that Tinnakorn protested. If he didn't do that, the water would have stayed longer.

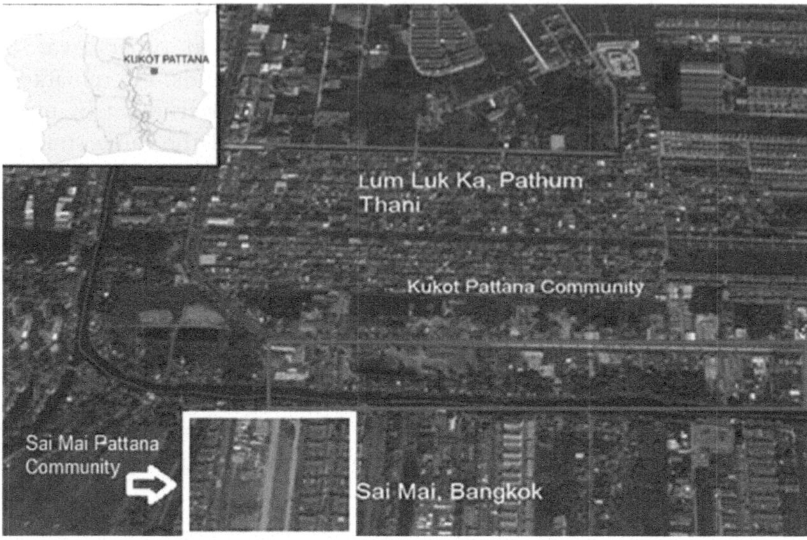

Fig. 10.2 Map of Kukot Pattana Community (*Source* Map from Google Earth under creative commons licence)

The protests helped improve the community.' Thus, this community contested the scalar configuration of the inner city remaining dry and the outer city being flooded.

Kukot Pattana Community: Contesting the Spatiality of Bangkok Staying Dry and Pathum Thani Being Flooded

The second case study community, Kukot Pattana, is located in the Kukot Municipal Area in Lam Luk Ka district of Pathum Thani Province (see Fig. 10.2).

A relatively new community, Kukot Pattana, is a slum community located along Khlong Hok Wa, which marks the boundary between Bangkok and Pathum Thani Province. In October, the floods reached up to around 2 metres from the ground level. Three actions taken by BMA blocked water from entering Bangkok from Lam Luk Ka but exacerbated the level of flooding in Kukot Pattana and other communities in Lam Luk Ka. First, BMA built a 2.5-metre-high sandbag wall in October—before the floods came—on the southern side of Hok Wa canal (see Fig. 10.2). The remnants of the wall can still be seen today. The wall prevented water from flowing out of the canal into Sai Mai. According to an ex-community leader in Sai Mai whose house was located on the southern side between the wall and the canal, the height of the wall was chest-level (at least one metre), and the flooding in his house, which was lower than the ground on which the sandbags were placed, was neck-high. His house as well as Kukot Pattana was flooded for three months. Second, Kukot Pattana residents inspected the nearby Khlong Song water gate which led into a sub-canal in Sai Mai. They found that the water level was two metres higher on the Lam Luk Ka side than on the Sai Mai side. Their finding clearly suggests that BMA was protecting Bangkok at Pathum Thani's expense. Third, the aforementioned big bag wall extended to Sai Mai and blocked water from flowing into the rest of Sai Mai from Lam Luk Ka. In particular, the big bags at the Directorate of Air Operations Control (DAOC) intersection along Phahon Yothin Road blocked water from flowing down this road which had transformed into a major waterway during the floods.

Residents of Kukot Pattana felt that BMA's actions were unfair. According to the community leader Gagae, because BMA was fully protecting Bangkok, people in his area were 'like floating ducks.' After the community had been flooded for a few weeks, 40 residents, plus residents from other communities, protested twice in mid-November at the DOAC intersection where the big bags were blocking the water from draining out of this area. During the first protest, they gathered and promised that they would remove the bags if FROC did not do so. FROC did nothing, so they protested again a few days later. During the second protest, the leaders of the protesters, including Gagae, demanded live on television (Channel 3) that FROC dismantle the entire barrier, stretching from Chulalongkorn sluice gate to Navanakorn Industrial Park in Pathum Thani so that water could drain out of their area (*Bangkok Post* 2011c). The *Bangkok Post* reported that while police were trying to negotiate with the protesters, some declared that they were tired of talking and removed sandbags from a section of a main barrier, causing floodwater to gush through the gap. The water level on both sides was deemed to be almost the same afterwards. Police threatened to arrest the protesters who responded that they were willing to be prosecuted. FROC's leader then promised that if the protesters dis-bursed, FROC would leave the 10-metre-wide breach open. The protesters agreed and disbanded (*Bangkok Post* 2011c). Kukot Pattana residents confirmed that the protest was successful in reducing the amount of water in their community. One of them who joined the protests remarked, 'When they removed the bags, the level of the water was lower.' She also confirmed that Gagae was one of the leaders of the protest. Another member who joined the protests said, 'After FROC removed the big bags, water level was still high but at least now we could use the main road to commute. The situation was better.' Gagae added, 'I saw that the level of water went down. After the big bag removal, water gradually reduced two centimetres per day.'

A few weeks later in November, Kukot Pattana residents organised again, this time protesting against BMA's management of the nearby water gate. Gagae and a few others had rode on a boat and investigated BMA's management of the gate. One resident stated, 'I went to see the water gate-BMA opened it only 20 cm-it would've been better if it was opened 30 cm. Also they were tricky: they opened it during the day and closed it at night.' Two days afterwards, a few dozens of community members gathered outside the water gate protesting against the unequal gate management and handed a petition to BMA and that requested its representative meet them to discuss the gate management. One of them said, 'Many of us went together and Khun Gagae went to talk to the BMA official. BMA agreed to widen the gate. The water drained faster after the opening of the water gate.' Gagae added:

> I wrote a letter asking BMA to open water gate from 10 cm to 40 cm. I sent a petition letter to the governor of BMA. The person who came to receive the letter was the head of Sai Mai district office. The agreement was to open the gate 40 cm –we weren't selfish, but we wanted to reduce water level … If BMA had a problem, they could tell people in the community and we would agree to reduce the height of the gate opening. When we protested, they would open the gate 40 cm and after we left, they would close it. When the media came they opened it … But after I met the representative and talked with him, BMA left it open … It was my own idea to open the water gate … I was born with the water and saw the way water flowed my entire life.

During both protests, Gagae used the media: 'I believed that media would be on the people's side because people were struggling. Every time I protested, I asked the media to join.' Since he worked a tow truck driver, he had collected the phone numbers of journalists working at television stations and newspapers when their vehicles had needed to be towed (he showed me their numbers as evidence). He then called these people the day before the protests and requested their presence.

Interviews of Kukot Pattana residents revealed that the majority of the interviewees (around 74%) either joined or supported the protests. Eight joined, two did not join because of their limited physical capacity but supported the protests, and seven said they would have joined if they had not evacuated. Six people said they neither joined nor supported the protest. Those who joined the protests gave the following reasons for doing so:

- "The water was smelly and there were many mosquitos so we gathered and protested. People were angry."
- "I went because I wanted to water level to be reduced. I wanted the big bags to be removed."
- "It was not fair-BMA blocked the water from not entering and not flooding that side [Sai Mai]. But here it was flooded badly. BMA should've let the water go."
- "I protested because of the water gate management-they didn't open the water gate so that's why amount of water was a lot. If they had opened the game, I wouldn't have protested at the big bag site."

Almost all of the interviewees blamed the Bangkok governor Sukhumbhand for making the flood worse and thought BMA did a poor job responding the floods. For example, Gagae stated:

> I think that the national government wanted Bangkok to share the water with other provinces. It wanted to let water flow through Bangkok but BMA refused. It did not want to collaborate. If you asked 100 people what happened, all of them would say that BMA tricked the national government.

Another resident commented:

> In the view of us people living outside Bangkok who were flooded, BMA was very bad. They didn't help us and made the flooding worse. If they shared the water, it would've been better. But they blocked the water using big bags instead of letting water go through.

These quotes illustrate that the protesters in Kukot Pattana had become infuriated with Sukhumbhand and BMA whom they felt were corrupt.

Thus, their protests were directed against BMA's flood management. These protests show that this community, along with others in Pathum Thani, had contested the configuration of dykes and water gates that caused Pathum Thani to be flooded while Bangkok remained dry.

Reasons for Numerous Protests During the 2011 Floods

The considerable number of protests raises the question of why so many protests arose during the floods. I suggest a number of answers. First, residents of Bangkok shared a widespread perception of government ineptitude in terms of its response to the flood. The findings of a poll conducted by Assumption University on 17–18 October 2011 showed that Bangkok residents gave FROC a mere 3.36 out of a total of 10 for its performance. Their overall confidence in the government also declined. Similarly, a poll conducted by Suan Dusit University surveyed 1357 people from 8 to 12 November and found that 64% of the respondents had less confidence in the government than before the floods began whereas only 5% had more confidence. Nationwide support for Prime Minister Yingluck also eroded as a result of her performance during the floods (Bangkok Pundit 2011). Another Suan Dusit University poll of more than 5000 Thais throughout the country revealed that their rating for Yingluck's performance dropped from 6.05 (out of 10) in September to 5.1 in October (Economist Intelligence Unit 2011). At the end of October, Kan Yuenyong, director of the think tank Siam Intelligence Unit, proclaimed: 'People now don't trust the leadership of Yingluck and the government' (Petty and Szep 2011).

Second, those outside the sandbag walls and closed water gates felt a deep sense of injustice. For example, 70% of interviewees in Don Muang and Lam Luk Ka Residents believed that the government's flood management was unfair. Inundated communities in the outskirts of the city felt that they were 'forgotten by authorities preoccupied with saving the shopping malls and skyscrapers of downtown Bangkok' (Wake 2011). Thus, many took a moral stance against the government's actions and became aggrieved.

Third, allegations of corruption fuelled this resentment. Corruption had weakened the country's relief efforts. 80% of respondents of a poll conducted by Assumption University believed that politicians were involved in corruption during the procurement of flood relief packages.

The Department of Special Investigation (DSI) alleged that the deputy director general of the Department of Disaster Prevention and Mitigation (DDPM) favoured two companies and that DDPM had bought relief bags at inflated prices (Ngamsaithong 2012). Likewise, residents of Kukot Pattana believed that BMA had purchased sandbags at inflated prices. The opposition Democrat Party also filed a no-confidence motion against the director of Flood Relief Operation Command who was the justice minister Pracha Promnok, accusing him of embezzling the state's flood aid budget and private donations (AFP 2011). Besides diminishing the amount and quality of the aid received by flood victims, corruption also lowered the public's perception of the government's response to the floods.

Fourth, rising political tensions and anger revolving around the country's political conflicts during 2005–2010 flared once again during the floods. As the conflict deepened, political mobilisation and polarisation occurred society-wide (Hewison 2015). Those who already opposed the government and voted against Yingluck Shinawatra and the Pheu Thai Party, and blamed her brother Thaksin for Thailand's failings, became further incensed when their areas were flooded, blaming the two siblings and the Pheu Thai Party. Similarly, those who had already opposed and voted against the opposition, ex-Prime Minister Abhisit Vejjajiva and his party, the Democrat Party, before the floods were further angered when their areas outside of Bangkok were flooded while many parts of Bangkok were dry or less flooded. These people felt that the Bangkok governor Sukhumbhand Paribatra, also a Democrat, protected the inner city of Bangkok and its majority of Democrat voters at the expense of other provinces, such as Pathum Thani, Nonthaburi and outer districts of Bangkok, such as Don Muang, all of which had more Pheu Thai supporters. For example, in Lam Luk Ka, a hotbed of Pheu Thai activists, many residents believed that BMA and Sukhumbhand had made the flooding worse. 62% of interviewees blamed BMA and the Democrats for the flooding. One resident of Kukot Pattana opined: 'The political conflict made it worse. BMA didn't want the water to enter Bangkok from here and here is a red-shirt area.'[2] She was one of the residents who joined both protests. Similarly, Kukot Pattana's community leader, who was also a Pheu Thai supporter and

[2]By red shirt, she meant Pheu Thai supporters. Numerous Thaksin supporters donned red shirts and protested against the Abhisit government in 2010.

who led the protests, felt that BMA had tricked the national government and was disappointed that Sukhumbhand was re-elected in 2013. In Yucharoen, one resident agreed: 'Sukhumbhand is not good because he only took care of downtown but not this area. I didn't vote for him (in the 2013 election).'

Fifth, protests have occurred in Thailand since the 1970s, but they became popular with a wide range of actors as a means of pursuing political goals during the Red-Yellow Conflict from 2005 to 2014. Most of the recent literature on protests in Bangkok focuses on mass protests by the opposing red shirts and yellow shirts which fought for political change at the national level (see McCargo 2010; Hewison 2014). After the Assembly of the Poor stopped mobilising in the late 1990s, mass protests were revived by the yellow-shirt rallies against Thaksin in 2005 and occurred from then until 2014. As Hewison argues, this period has been one of 'essentially nonstop protest' in which both sides of the political conflict have developed a 'reliance on street-based politics' (2014, pp. 1 and 4). While certainly some grassroots groups joined these protests, as McCargo contends (2010, p. 8), the conflict has been at the national level 'between different elements of the Thai elite, who have mobilised rival patronage-based networks of supporters' to protest. These protests occurred mostly within central Bangkok, such as in Ratchadamnoen, home to traditional political arenas, and Ratchaprasong, home of luxury shopping malls and elite urban space from which the urban and rural poor have been excluded. By protesting at Ratchaprasong, those who have been excluded from this space showed that they were unwilling to passively remain marginalised and have their political voice muted (Vorng 2012). A number of community leaders drew upon this recent collective memory of protests during the flooding as a means to contest the government's flood management. These five reasons combined during the floods to create an environment conducive to activism at the urban level of Bangkok.

Urban Activism in Bangkok: Contesting the State's Governance of Flooding

During the 2011 floods, many inhabitants in Bangkok turned to collective action after becoming aggrieved by the state's inept response, the inequalities in terms of exposure to the floods and corruption allegations.

Comparing case studies, a number of similarities emerge. A main ingredient of these protests was a charismatic and authoritative leader who initiated and organised the protests. In both cases, community leaders became aggrieved about the unfairness of the state's response, either in terms of floodwater distribution or compensation distribution, and rallied their members to join them in protest. These community leaders had gained authority within their communities because they had been elected and the majority of community members approved of their leadership. For example, 96% of interviewees believed that Gage was a good leader.

During the flooding, community leaders' connections were important in two ways. The media was a key tool used by protest leaders, particularly Tinnakorn and Gagae. Both used their connections to request that members of the media to cover both protests and subsequent negotiations with government officials. Newspaper journalists and TV reporters agreed to these requests. In both instances, government leaders yielded to their demands. The pressure from being live on television or in front of newspaper reporters likely played a major role in prompting them to comply. This also suggests that government leaders felt they had a degree of downward accountability to those who were badly flooded and did not want to appear negative in front of the public. Moreover, the community leaders' connections to other nearby community leaders also helped them mobilise others to join the protests.

Also, the degree of legitimacy of the negotiators' argument affected their bargaining power. This was especially the case in Yucharoen. By drawing upon constitutional and human rights and calling themselves quality citizens of Bangkok, Yucharoen residents were able to pressure FROC to comply with their demands. During the second round of negotiations, when BMA used economic logic, asking Yucharoen residents to sacrifice their area for a few more days in order to protect the economy of Bangkok, which, as BMA argued, was the heart of the Thai economy, the two sides compromised.

Third, these protests were fuelled by the memory of recent protests against inequities during the previous years which had resulted in political changes and by the country's deep-seated political polarisation. Led by their community leaders, protesters sought to disband sandbag walls and to open water gates, both of which blocked water from leaving their communities. They employed a number of strategies, such as making symbolic votes, signing petitions and organising mass gatherings. In particular, they sought to challenge the neoliberal logic followed by both FROC and BMA that the 'inalienable rights of individuals to private property and the profit rate trump any other conception of inalienable rights you can think of' (Harvey 2009, p. 43). This ideology was demonstrated when FROC and BMA protected the inner city and industrial estates at all costs.

Fourth, similar to the 2010 Red Shirt protests, the two 2011 case studies demonstrate, while the protesters certainly succeeded in pushing the state to change its actions, their degree of success was limited. For example, the state only allowed gaps in the walls, not an entire dismantling, or only opened the water gates somewhat, not fully. The state still protected the inner city the most and created uneven vulnerabilities to the floods. Nonetheless, during the floods, the high number of protests still reduced the vulnerability of protesters, altered the spatiality of the floods and improved their mental state.

As in most conflicts from 2005 onward in Thailand, the political polarisation between the two sides (red and yellow) pervades these disputes as well. For example, as interviewees stated, Sukhumbhand did not trust the protesters in Don Muang and Lam Luk Ka because he was a Democrat, and they were Pheu Thai supporters. In addition to their strong sense of injustice, residents in Lam Luk Ka also protested because of this difference, believing that Sukhumbhand had tricked the national government which these residents supported. Thus, the governance of the floods became yet another event which polarised the country between the two different sides and sparked protest.

Conclusion

These small and limited triumphs emphasised that urban protests were largely local and disparate, aimed at addressing local flooding, not wider issues of rights and justice. Nonetheless, the geography of protest spread to new suburban areas on the boundaries of the floodwater in northern, northwest and northeast Bangkok, modelled in part on the memory of recent collective political protests, including the 'Red Shirt' occupation of Central Bangkok in anti-government protests (see Vorng 2012). Their frustrations at Central Bangkok being completely dry while their peripheral areas remained heavily flooded for weeks were instrumental in their mobilisation against FROC's and BMA's unequal governance of the floods. Thus, the recent history of political dissidence provided useful for anti-government flood protests.

Further, these peripheral areas thus became new spaces of mobilisation against FROC's and BMA's unequal governance of the floods. In hitherto quiescent spaces and places, protesters constituted, much as in Brazil, 'movements of insurgent citizenship to confront the entrenched regimes of citizen inequality that the urban centres use[d] to segregate them' (Holston 2009, p. 245). Yucharoen protesters repeatedly asserted: 'We are quality citizens of Bangkok too' while protesters in Phrom Samrit argued: 'People should be treated equally.' Much as members of slum communities in Khon Kaen protested and Red Shirts in Ratchaprasong in 2010 because they aspired to participate equitably in Thai society (Elinoff 2012), suburban citizens of Bangkok demanded a more inclusive notion of citizenship, shaped and articulated during the floods, by demanding equal rights and equal access to dry space. By seeking to reshape the uneven spatiality and imposed spatial injustice of the disaster, they collectively sought to correct injustices, remake urban space and claim their right to the city.

The Bangkok context demonstrates the strong linkage between flood justice and rights to the city. During the 2011 floods, Bangkok residents demanded a more inclusive notion of citizenship and achieved limited and localised gains. In response to injustice and the denial of equitable benefits from urban life, through the unfair distribution of environmental harms, which built on socioeconomic inequalities, they collectively fought for a redistribution of harms and benefits, as Lefebvre (1993) argued should happen. Consequently, as Douglass argues, 'hope must also be placed on these projects big and small to reclaim the city and its spaces' (2014, p. 27) because they show the way forward.

References

AFP. (2011, November 27). *Thai Minister Facing No Confidence Vote Over Flood.*

Bangkok Metropolitan Administration. (2012). *Statistical Profile of Bangkok Metropolitan Administration 2011.* Bangkok: Bangkok Metropolitan Administration.

Bangkok Post. (2011a, November 12). Govt Apologises to Flood Victims.

Bangkok Post. (2011b, November 14). Don Muang Bags Protest Grows.

Bangkok Post. (2011c, November 16). Big Bag Barrier Breached Again.

Bangkok Post. (2011d, December 12). Don Muang Should Be Dry in 3–5 Days. http://www.bangkokpost.com/archive/don-muang-should-be-dry-in-3-5-days/270336.

Bangkok Pundit. (2011). *Suan Dusit Polls Show Declining Confidence in Thai PM, Govt.* http://asiancorrespondent.com/2011/11/suan-dusit-polls-show-declining-confidence-in-pm-and-the-thai-government/.

Collins, T. W. (2010). Marginalization, Facilitation, and the Production of Unequal Risk: The 2006 Paso Del Norte Floods. *Antipode, 42*(2), 258–288.
Dikec, M. (2009). Justice and the Spatial Imagination. In P. Marcuse, J. Connolly, J. Novy, I. Olivo, C. Potter, & J. Steil (Eds.), *Searching for the Just City: Debates in Urban Theory and Practice* (pp. 72–88). London and New York: Routledge.
Douglass, M. (2014). After the Revolution: From Insurgencies to Social Projects to Recover the Public City in East and Southeast Asia. *International Development Planning Review, 36*(1), 15–32. https://doi.org/10.3828/idpr.2014.2.
Economist Intelligence Unit. (2011). *Floods Present a Serious Test of Yingluck's Leadership.*
Elinoff, E. (2012). Smouldering Aspirations: Burning Buildings and the Politics of Belonging in Contemporary Isan. *South East Asia Research, 20*(3), 381–398. https://doi.org/10.5367/sear.2012.0111.
Elinoff, E. (2013). *Architectures of Citizenship: Democracy, Development, and the Politics of Participation in Northeastern Thailand's Railway Communities.* PhD dissertation, University of California, San Diego. http://escholarship.org/uc/item/6x1444bm.
Frederickson, T. (2011, November). Flood Management Controversy. *Bangkok Post.*
Harvey, D. (1973). *Social Justice and the City.* Athens: University of Georgia Press.
Harvey, D. (2009). The Right to the Just City. In P. Marcuse, J. Connolly, J. Novy, I. Olivo, C. Potter, & J. Steil (Eds.), *Searching for the Just City: Debates in Urban Theory and Practice* (pp. 40–51). Abingdon, Oxon and New York, NY: Routledge.
Hewison, K. (2014). Thailand: The Lessons of Protest. *Journal of Critical Perspectives on Asia, 50*(1), 1–15.
Hewison, K. (2015). Thailand: Contestation Over Elections. *Sovereignty and Representation, 51*(1), 51–62. https://doi.org/10.1080/00344893.2015.1011459.
Holston, J. (2009). Insurgent Citizenship in an Era of Global Urban Peripheries. *City & Society, 21*(2), 245–267.
Jain, H. (2010). Community Protests in South Africa: Trends, Analysis and Explanations (Local Government Working Paper Series No. 1). University of Western Cape. Community Law Centre. http://mlgi.org.za/publications/publications-by-theme/local-government-in-south-africa/community-protests/Final%20Report%20-%20Community%20Protests%20in%20South%20Africa.pdf.
Keil, R. (2003). Urban Political Ecology1. *Urban Geography, 24*(8), 723–738.
Krungthep Turakij. (2011, November 17). *Chao Lum Luk Ka Ruu Nao Gan Nam Tuam Khet Sai Mai* [Lam Luk Ka People Removes the Water Barrier. Floods in Sai Mai District].

Lefebvre, H. (1993). *Architecture Culture: 1943–1968* (J. Ockman, Ed.). New York: Rizzoli.

Lefebvre, H. (1996). *Writings on Cities* (E. Kofman & E. Lebas, Eds. and trans.). Oxford: Blackwell.

MacCallum, D., Steele, W., Byrne, J., Houston, D., & Others. (2011, November). *Environmental Imaginaries: Climate Change as an Object of Urban Governance*. State of Australian Cities, Melbourne.

Marcuse, P. (2012). Whose Right(s) to What City? In N. Brenner, P. Marcuse, & M. Mayer (Eds.), *Cities for People, Not for Profit: Criticual Urban Theory and the Right to the City* (pp. 24–41). Abingdon, Oxon and New York, NY: Routledge.

Marks, D. (2015). The Urban Political Ecology of the 2011 Floods in Bangkok: The Creation of Uneven Vulnerabilities. *Pacific Affairs, 88*(3), 623–651.

Marks, D., & Lebel, L. (2016, March). Disaster Governance and the Scalar Politics of Incomplete Decentralization: Fragmented and Contested Responses to the 2011 Floods in Central Thailand [Decentralizing Disaster Governance Special Issue]. *Habitat International, 52*, 57–66.

McCargo, D. (2010). Thailand's Twin Fires. *Survival, 52*(4), 5–12. https://doi.org/10.1080/00396338.2010.506815.

Murray, M. J. (2009). Fire and Ice: Unnatural Disasters and the Disposable Urban Poor in Post-Apartheid Johannesburg. *International Journal of Urban and Regional Research, 33*(1), 165–192.

Ngamsaithong, N. (2012, January 9). DSI Finds Corruption in Flood-Relief Bag Procurement. *Thai Financial Post*. http://thaifinancialpost.com/2012/01/09/dsi-finds-corruption-in-flood-relief-bag-procurement/.

Ortmann, S. (2015). The Umbrella Movement and Hong Kong's Protracted Democratization Process. *Asian Affairs, 46*(1), 32–50. https://doi.org/10.1080/03068374.2014.994957.

Pelling, M., & Dill, K. (2009, May). Disaster Politics: Tipping Points for Change in the Adaptation of Sociopolitical Regimes. *Progress in Human Geography*. https://doi.org/10.1177/0309132509105004.

Petty, M., & Szep, J. (2011, October 28). Insight: Thai Flood Crisis Puts Swamped PM in Firing Line. *Reuters*. http://www.reuters.com/article/us-thailand-floods-insight-idUSTRE79R0NK20111028.

Roth, D., & Warner, J. (2007). Flood Risk, Uncertainty and Changing River Protection Policy in the Netherlands: The Case of 'Calamity Polders'. *Tijdschrift Voor Economische En Sociale Geografie, 98*(4), 519–525.

Schlosberg, D. (2007). *Defining Environmental Justice: Theories, Movements, and Nature*. New York: Oxford University Press.

Sheppard, E., Gidwani, V., Goldman, M., Leitner, H., Roy, A., & Maringanti, A. (2015). Introduction: Urban Revolutions in the Age of Global Urbanism. *Urban Studies, 52*(11), 1947–1961. https://doi.org/10.1177/0042098015590050.

Sökefeld, M. (2012). The Attabad Landslide and the Politics of Disaster in Gojal, Gilgit-Baltistan. In U. Luig (Ed.), *Negotiating Disasters: Politics, Representations, Meanings* (pp. 175–204). Frankfurt am Main: Peter Lang Verlag.

The Nation. (2011, November 25). Bang Khae Residents Block Expressway. http://www.nationmultimedia.com/national/Bang-Khae-residents-block-expressway-30170601.html.

Vorng, S. (2012). *Incendiary Central: The Spatial Politics of the May 2010 Street Demonstrations in Bangkok* (MMG Working Paper 12–04). Max Planck Institute for the Study of Religious and Ethnic Diversity, Göttingen.

Wake, D. (2011, November 27). Misery Lingers for Bangkok's 'Forgotten' Flood Victims. *Agence France-Presse*. http://www.abs-cbnnews.com/global-filipino/world/11/27/11/misery-lingers-bangkoks-forgotten-flood-victims.

Walker, D. M. (2013). Resisting the Neoliberalization of Space in Mexico City. *Locating Right to the City in the Global South, 43*, 171–194.

Wong, N. W. M. (2016). Environmental Protests and NIMBY Activism: Local Politics and Waste Management in Beijing and Guangzhou. *China Information, 30*(2), 143–164. https://doi.org/10.1177/0920203X16641550.

CHAPTER 11

The Evolution of Housing Rights Activism in South Korea

Seon Young Lee

Introduction

Even though the concept of housing rights has been gaining popularity among policy-makers and even ordinary people in South Korea (hereafter Korea), housing rights have various meanings on both a social and an individual level. Leckie (1992, p. 4) defined housing rights in two ways. One definition relates to the actual characteristics of a house: 'its physical structure, the cost, location, and the infrastructure facilities supplied to it, the site on which it is built and legal security of tenure'. These are not rights, they are characteristics, but security of tenure is a right. This definition is similar to outlining basic needs. In the Korean context, it is important to take into consideration Leckie's second definition of housing rights, which is 'the right to public participation and the right to determine one's own destiny in housing matters' (Leckie 1992, p. 5). Furthermore, the notion of housing as a social right is still unfamiliar, since housing has been treated as a commodity in the market under the family-based welfare system; it has operated without sufficient support

S. Y. Lee (✉)
Independent Researcher, Seoul, South Korea

N. M. Yip et al. (eds.),
Contested Cities and Urban Activism, The Contemporary City,
https://doi.org/10.1007/978-981-13-1730-9_11

from the state for a long time (Kim 2006b). There was no comprehensive welfare system under the 1970s and 1980s authoritarian developmental state in Korea, and the state did not take any responsibility for housing at that time (Woo 2004, p. 203). Lee (1999, p. 26) argued that

> For the authoritarian developmental state, the welfare component was to be kept marginal and legitimised only by the national goal of economic development, which is seen as the only ground for authoritarian rule.

The broad welfare system that was eventually introduced was based on 'productivist welfare capitalism':

> In the Korean case of productivist welfare capitalism, there is a heavy reliance on private sector delivery. Where delivery is undertaken by public-sector agencies, it is usually accompanied by high user fees ... The main objective of the social insurance system is to benefit productive sectors of the economy and to contribute to the smooth operation of the labour market ... One of its key functions was to ensure the suppression of labour activism and the direction of national energies toward productive activity. Consequently, only a very small proportion of government expenditure was spent on social development. (Kwon and Holliday 2007, pp. 243–244)

Therefore, social services tend to be provided in terms of a selective social welfare perspective and, indeed, housing policy has also been implemented in this way. The state provides housing allowances and public rental housing for the least well off, but there have not been sufficient efforts made to establish rent control, security of tenure and housing management for the general public. However, people's interest in the universal welfare system has been growing. As more citizens have become interested in social welfare, the state now feels pressed to take on greater responsibility for society as a whole.

In reflecting on this change, this research aims to develop a greater understanding of housing rights activism in Korea over the last four decades. The next section reviews the literature on social protests around housing and urban development with regard to urban activism in the West and in East Asia. It then outlines a typology of urban social movements (USMs) through the creation of a comparative general framework in order to shed light on urban activism in Korea. The third section sets out the comprehensive background of housing policies. The fourth section sheds light on the nature, form and outcomes of housing rights activism by members of the civil society that challenges the dominant social power relations. The final section summarises the research findings.

Urban Conflicts and Social Changes

The city is a social product resulting from conflicting social interests and values (Castells 1983, p. 291).

Castells (1976, p. 155) introduced the term 'urban social movements' (USMs) to describe collective citizen action designed to bring about structural social changes. USMs result from the uneasy relationship between capitalism, democracy and participation in the urban system. Castells (1983) argues that mobilisation can transform this urban structure, and this is why he calls this change of urban power relations a USM. In order to achieve change, people have mobilised themselves—this has been the case from the US civil rights movements in the 1960s through to contemporary movements like the Occupy Wall Street movement. USMs have changed in terms of their goals, mobilisation methods and characteristics since Castells first investigated them (Mayer 2006; Rabrenovic 2008). The rise and fall of USMs must be understood in terms of the broader political and economic contexts that lie beyond the urban level (Fainstein 1985). Even though USMs have evolved through changes to historical conditions, USMs represent citizens mobilising themselves in reaction to social conflicts and using protests to prompt a solution to their problems to be found. USMs are one of the few means of expression for people who do not have the opportunity to access sociopolitical resources. They have played an important role in resolving the problems faced by such groups. In particular, USMs are crucial challenges to urban politics, which has lost its ability to meet citizens' demands.

In East Asia, it is not easy to make social movements successful. As the East Asian economy has developed rapidly, property has become highly valued. This has induced social conflicts over property development and profit distribution. The strong states have pushed urban (re)development and housing policies in the name of the common good, but the policies have triggered gentrification and segregation. There is no input from the public in top-down urban redevelopment practices, and residents are only contacted when urban (re)development programmes occur in their area (Ng 2002; Shin 2008). Large-scale displacement and eviction over urban (re)development are more common and direct in East Asian gentrification (Wu 2004; He 2012). Mega-events and large-scale development have caused the loss of affordable housing and detrimentally affected the lives of the urban poor (Shin 2009; Davis 2011). However, urban activism for housing rights has not been easy to develop due to the strong state and weak civil society. Nevertheless, resistance against eviction and displacement has been occurring in many East Asian cities (e.g., Kim 1998; Wu 2004; Smart 2009; Weinstein 2009; Shao 2013). Some protesters have acquired better compensation and relocation sites and have hindered redevelopment projects by stopping developers (Shin 2013). This urban activism has made progress in advancing protestors' housing rights in terms of legal procedures. Residents have become active in exercising their rights in the face of legislation that allows forced demolition and eviction.

However, the success or failure of USMs is highly dependent on whether or not they are part of broader political movements. McGovern (1997, p. 421) emphasised that

> Political change is the key to breaking out of the conventional pattern of limited urban policy making that generally benefits the interests of the downtown business community and its allies.

Political changes are important in making social protests successful, but Tarrow (2011, p. 6) stresses the importance of social protests in political changes. He argues that collective grass-roots movements could trigger important political changes, even if they fail. Social protests have changed urban policies, and urban policies have changed the characteristics of social movements. Some of them achieved their goals, but others failed or lost their momentum. However, USMs could be the most powerful form of collective opposition to socio-spatial development. USMs are diverse in each country and are not identical even in a single country over time. There is a need for a structure to allow for better comparisons and contrasts between USMs. Pickvance (1985, p. 44) suggested five contextual features that affect the incidence of USMs, as shown in Table 11.1. This is useful for improving our understanding of how and why each USM occurred when it did and in which countries, and is helpful when accounting for the outcomes of USMs.

Pickvance (1985, p. 39) outlined a typology of USMs in terms of collective consumption, local-level political process and spatial proximity,

Table 11.1 Five contextual features affecting the incidence of urban social movement

Dimensions	
Rapid urbanisation	Yes/no
State action	Towards movements (tolerant/intolerant) Intervention on consumption (yes/no)
Political context	Presence of broad political mobilisation (yes/no) Cultural meaning of urban politics (class-based or not) Effectiveness of political institutions in expressing political conflicts (no opposition parties/no alternation/alternation)
Development of middle class	Yes/no
General economic and social conditions	Encourage/discourage protest

Source Pickvance (1985, p. 44)

Table 11.2 Pickvance's typology of urban social movement

Category	*Goal*
Type 1	Movements regarding the provision of housing and urban services
Type 2	Movements over access to housing and urban services
Type 3	Movements regarding control and management
Type 4	Defensive movements against physical threats

Source Pickvance (1985, p. 44)

as shown in Table 11.2. To an extent, Pickvance's typology is unable to categorise USMs clearly since there is some overlap between the categories—something Pickvance notes himself. However, this typology is useful for comparative analysis, since it can take into account the causes, circumstances and militancy of mobilisation. In the context of USMs in Korea, Pickvance's framework may not fit well since Korea has a very different cultural and political history from the West. However, it could be useful when comparing the Western context to the Korean context. It could also suggest how USMs in Korea have evolved as they have caught up with more advanced countries. The framework can additionally provide an insight into the evolution of urban activism in Korea.

With regard to the rise and fall of the USM in Korea in relation to Pickvance's framework, it can be said that the state could not keep pace with the rapid urbanisation, and the provision of collective consumption was not a high priority in state policy. The working and middle classes had limited power during the industrialisation period, whereas the authoritarian state had strong control over society and was intolerant of USMs. Social protests were thus brutally repressed for the purpose of economic growth under authoritarian and military regimes. It was hard to initiate USMs. The goal of social movements in Korea was to build a liberal and democratic state instead of an authoritarian one. However, citizens in Korea have recently been able to access more politically representative systems such as local electoral ones. Political conflicts have been more effectively expressed, and state action against social movements has been more tolerant and flexible.

Housing Policies and Homeownership Developmentalism in Korea

Housing is essential to all, so the housing issue should be approached in terms of the proper functioning of society. However, housing developments in Korea have only been covered by relatively liberal policies when compared to the state-regulated urban development policy (Doling 1999). The state gave less priority to housing in its developmental strategy (Kim 2010). Housing policies have been subordinate to economic policies and pricing strategies, so they have not met the needs of low-income households (Lim 2005, p. 325). As housing and land policies have been dealt with as commodities rather than social goods, property speculation has created opportunities for mobility and class differentiation and social inequality has increased. Shin (2009, p. 907) also notes that 'housing provision in South Korea has been market-oriented with minimum involvement of the state to address urban poor families' housing needs'. This is due to the country's predominant economic policies and the nature of the state, which is dependent on a growth alliance with 'chaebols' (business conglomerates). Shin (2011) coined the term 'homeownership developmentalism' to explain these characteristics of housing development in Korea. He argues that Korean residential capitalism has a unique structure. The number of owner-occupiers and the level of mortgage debt to GDP are both low, but there is not enough public input into the consumption and production of housing. Because of the lack of public intervention in housing policy, the private sector (e.g., construction and investment companies) and those social classes with purchasing power have controlled the housing market (Shin 2011, p. 132).

However, when political crises emerged, the central government encouraged the growth of housing supply and seriously considered housing policies for the stability of the regime (Lim 1995). Korean governments dealt with housing policy on an ad hoc basis according to short-term political considerations. Housing policy was viewed as a part of industrial policy, so the state encouraged housing production but this was not for the reproduction of labour power (Yoon 1994, p. 84). The Korean state did not play an important role in the provision of housing, but established quasi-governmental companies to increase housing provision. The profit-orientated private sector has predominantly constructed housing, and this has influenced housing commodification (Lee 2003). The government has implemented housing policies in favour of housing suppliers in order to increase housing provision and economic growth and avoid the use of public funds (Ronald 2010). Housing policy has focused on boosting homeownership and meeting middle-class demand. Housing consumption is constrained by individual households' economic ability to pay, since the Korean state focuses on a filtering-down strategy rather than direct interventions like rent control or social housing provision (Ha 2002).

By relying heavily on private investment to meet the housing demands, there is not enough social housing for poor or middle-class households. Although the state has not provided reasonable policies to promote housing provision, the urban population has grown rapidly. The increasing demand for housing and the inadequate supply have caused large increases in house prices and property speculation. Accordingly, property speculation has become a profitable activity and this has resulted in social differentiation (Jang 2006). The state has continuously made attempts to expand homeownership, but the homeownership rate has declined since the 1950s. Widespread property speculation has caused lower- and even middle-income households to experience difficulties in becoming homeowners. The state has implemented various anti-speculation measures, but these have not been effective in tackling speculation (Ronald 2010).

In fact, the state has stimulated speculation on land and housing in order to boost the economy. Development plans and construction works have been used as important growth machines, since construction has been considered the most effective industry for increasing jobs and boosting the economy due to its forward and backward linkage effects. As a result, a belief in the idea that property value never decreases emerged and became widely believed. Over the last four decades, many people have come to favour property investment or speculation. Property speculation in search of profit has become a 'national hobby'. It has continuously increased the property bubble and created property classes.

In summary, housing policies in Korea can be explained as a key component of Korean developmentalism. Housing policies have suffered from the lack of a long-term vision. The state has focused on housing provision for homeownership expansion, so there have been insufficient policies to help low-income households and to supply affordable housing or housing benefits for general tenants. Property speculation has been a chronic problem, but the state has not effectively dealt with it since the property market and economic growth have relied on each other. The state has implemented a housing policy as one of its economic policies and focused on housing provision for middle-class households based on the housing filtering theory. Therefore, Koreans have been forced either to buy a flat or deal with highly insecure private tenancy and pay high private rents. Private landlords have not taken social responsibility, but there has been no alternative to the rental market since social housing and housing associations do not exist in Korea.

Seeking Alternatives at the Grass-Roots Level

The state has tried several urban redevelopment policies to solve the housing shortage, and urban redevelopment has been the main tool used for upgrading residential conditions and housing supply. However, urban redevelopment has strengthened the commodification of space and housing, the violation of property and housing rights, and social polarisation. Wholesale clearance and demolition have been met with opposition from residents in urban redevelopment areas, but many urban redevelopment projects—with very few exceptions—have been completed after suppressing neighbourhood resistance. Forced eviction and displacement have become widespread and taken place systematically in the name of urban redevelopment. As a number of people and communities have been affected, more social protests and resistance have emerged. However, there have been insufficient channels through which to halt urban redevelopment that is against the will of residents. Urban redevelopment has induced and strengthened social problems and led to housing rights activism. The evolution of housing rights activism in Korea can be explained by changes in urban redevelopment policies.

Under the strong state and relatively weak civil society prevalent during the 1960s and 1970s, urban redevelopment programmes caused large-scale eviction and displacement. In spite of this, it was not possible for evictees to form collective movements. Civil society was not yet developed under the authoritarian rule imposed by the military regime (Kim 2003, p. 196). Social protests were severely suppressed by the authoritarian state. Therefore, it was difficult for anti-eviction movements or housing rights activism to collectively break out and succeed. However, more resistance and protests against eviction and displacement broke out in the 1980s. Social protests over housing issues since the 1980s can be understood in terms of the emergence and development of urban activism for fundamental social change, not just activism that aim to promote evictees' rights to housing. This evolution should be interpreted through consideration of the relationship between the transformation of broad urban activism and political environments. In general, the state was intolerant of anti-urban redevelopment movements and politically representative systems (such as local electoral systems and opposition parties) until the late 1980s. The state's actions have become more tolerant and flexible, and political conflicts have been more effectively expressed since the early 1990s. Housing rights activism has become more diverse in terms of its key actors, the form it takes and its relationship with other movements and the state. The transformation of urban activism over housing in Korea can be divided into three waves, as shown in Table 11.3.

The First Wave of Housing Rights Activism

The first two waves can be connected to the Type 4 defensive movement against physical threats suggested in Pickvance's typology, which is a direct response to demolition, eviction and financial losses as suffered by protesters in a crisis situation. After the introduction of the Joint Redevelopment Project policy (hereafter JRP) in 1983, the first wave of housing rights activism occurred. Property owners and construction companies became the main actors in the process of eviction and demolition, and the state did not directly take part in the process. This resulted in a new conflict relationship: tenants vs property owners. In the past, social conflict over urban redevelopment involved the state and low-income groups; however, this changed to social conflict between interest groups in the 1980s. The local state did not take action to resolve

Table 11.3 Transformation of housing rights activism in Korea

	The first wave	*The second wave*	*The third wave*
Period	1983–2002	2003–2012	2012–present
Conflict relationship	Tenants vs landlords Tenants vs the state	Tenants vs landlords Tenants vs the state Owner-occupiers vs the state Owner-occupiers vs absentee landlords	Non-homeowners* vs the state Non-homeowners vs the market
Goals	Keep basic needs and housing rights Social justice	Keep basic needs and housing rights Keep property rights Social justice Control and management of urban redevelopment	Keep basic needs and housing rights Social justice Control and management of housing and urban planning
Strategies	Collective social protests Establish evictees' own centralised social organisations Support from religious and student groups	Collective social protests Establish owner-occupiers' own centralised organisations Form relationships with political parties Lobby politicians and litigate	Collective social protests Establish tenants' own centralised organisations Form relationships with political parties Support from various social institutions
Outcomes	Introduce compensation packages and social housing Improve evictees' rights	Halt urban redevelopment Introduce new urban redevelopment policies	Change housing paradigm Introduce new housing policies Introduce new housing supply methods
Pickvance's typology	Type 4 Type 1 Type 2	Type 4 Type 2 Type 3	Type 1 Type 2 Type 3

Note 'Non-homeowners' mean those who do not own their home this conflict because the problems induced by the JRP were considered the results of a private conflict between the property owners' association and tenants (Hwang 1989). Even though the poor made claims to their local authorities, the lack of a local election system meant protesters lacked a political connection to their local mayor or opposition parties (Lee 1990). Since the JRP policy was enacted on a large scale, many people across Korea were affected. Accordingly, many tenants took collective action against property owners and construction companies.

Anti-eviction movements received support from democracy movements and labour movements in the late 1980s (Kim 1998). Religious groups also started to actively help the mobilisation of the urban poor's anti-eviction movements and served as a shield that protected the urban poor from brutal suppression by the state (Kim 1998, p. 241). Religious groups and students' groups contributed to strengthening anti-eviction movements by helping people to set up meetings and organising committees. Protesters resisted forced demolition and violent eviction through marches, street demonstrations, vigils and hunger strikes. They fought ferociously and often clashed with police and demolition thugs. When they did not succeed in achieving any fundamental changes, some protestors chose suicide as an extreme form of protest. According to research by Kim et al. (1998), 11 protesters committed suicide between 1986 and 1998.

The first wave is distinct from the 1970s anti-eviction movements, which focused on housing and the basic necessities of life. Tenants started to realise the social injustice and unfairness of urban redevelopment, which excluded them from the economic gains of such projects. People started to demand just law enforcement and housing rights from the state (Kim 1998), while tenants demanded just compensation from the state and landlords, such as alternative housing (Cho 1989). Urban activism over urban redevelopment began to expand beyond calls for compensation and anti-eviction movements. They started to demand social welfare systems and greater governmental responsibility regarding urban redevelopment policies and the lack of affordable housing. In addition, a new trend appeared in terms of the key actors and mobilisation methods. While previous activism in the 1970s was sporadic, protesters in the 1980s began to resist redevelopment continuously. They started to collaborate with other people facing these problems. As a result, the first organisation for evictees—the Seoul Council of the Centre for Victims of Forced Evictions (the SCFE)—was established in

1987 by victims of eviction. Social protests against eviction and displacement became more systemised. Tenants facing similar problems gathered to fight for their housing rights together with others from urban redevelopment areas all around Seoul. The establishment of the SCFE enabled people to access institutional support for protest activities in each neighbourhood and facilitated the growth of active anti-eviction movements and housing rights activism.

Tenants' movements during this period were successful at promoting public awareness, and they made the issues they raised political problems. These changes can be interpreted in the light of the collapse of the authoritarian regime and the development of a civil society out of the political democratisation movement of the 1980s (Kim 2006a). These strong social protests led to some crucial transformations to housing policies, and a new compensation system was introduced to improve the urban poor's and tenants' housing rights. Since then, the state has become more involved with housing provision for the low-income groups and has assumed responsibility for the protection of their housing rights (Ha 2002). In 1989, the president announced that the state would supply affordable social housing and permanent-rental social housing (Ha 2002). In 1991, the mayor of Seoul settled an agreement regarding temporary accommodation and the compulsory building of social housing for rent in urban redevelopment areas. These changes meant that tenants would be eligible for compensation in the redevelopment process. A new urban redevelopment policy called the Urban Poor's Housing Environment Improvement Act was introduced in 1989 to improve the living conditions of low-income groups, and the state took more responsibility for urban redevelopment (Lee 2000). This was a great step forward, but this progress was not fully realised due to the characteristics of the property-led accumulation coalition (Lee 2003). Housing rights were still not guaranteed, and the protection the state offered was inadequate (Kim 1998, p. 248). Therefore, social resistance against urban restructuring continued and more social organisations were established during the 1990s. New social organisations such as the Korea Coalition for Housing Rights established in the 1990s moved beyond focusing on anti-demolition issues and concerned themselves with housing rights and social welfare. These efforts led to the introduction of the Housing Act 2003, which regulated housing standards. Housing started to be considered a right rather than a need (Park and Lee 2012). In short, the first wave led to state intervention in housing provision and consumption regarding the poor, and, after the state took action by increasing housing provision, activists in this wave went on to push for access to this housing. The first wave of activism developed into Pickvance's Type 1 and Type 2 movements from what was a Type 4 movement.

The Second Wave of Housing Rights Activism

Large-scale anti-eviction and urban redevelopment movements re-emerged in the late 2000s. New anti-urban redevelopment movements, which are termed owner-occupiers opposition movements, emerged after a new urban redevelopment policy named the New Town Project was launched in 2003 (Kim 2010). Owner-occupiers activism became more active and visible after the 2008 economic recession (Lee 2016). This is quite a new form of social activism against urban redevelopment in Korea, so it is under-researched, with only a little empirical research having been conducted on the subject (e.g., Kim 2010; Shin 2010; Lee 2014). There is no substantial literature that examines the effects of these movements on the urban redevelopment process. However, owner-occupiers activism has influenced changes in the state's approach to urban redevelopment.

Owner-occupiers were relatively passive in previous anti-urban redevelopment movements, but they actively mobilised in the second wave. In general, owner-occupiers were in favour of urban redevelopment because it enabled them to increase their property's value; by obtaining a new flat, they secured themselves a middle-class status symbol and a good investment (Park 1988; Lee 2013b). However, owner-occupiers went on to become major actors in resistance to urban redevelopment, making tenants' movements appear relatively weak by comparison. In the second wave, more conflict relations were produced between different interest groups. Social conflict over urban redevelopment became more complicated and expanded along with existing conflicts between tenants and other parties. However, it was not easy for owner-occupiers to organise their struggles in the beginning, and they were not given enough attention and support from social organisations involved with housing rights and anti-eviction movements. Existing social organisations were reluctant to get involved with this new form of activism, since owner-occupiers were seen as stakeholders who stood to gain economic benefits from urban redevelopment (Kim 2010, p. 169). Owner-occupiers have also kept away from existing anti-urban redevelopment organisations. Therefore, owner-occupiers working against urban redevelopment found it hard to establish their credibility with other social organisations. As a result, they organised the Nationwide Coalition of Property Owners for Immediate Counteraction to Urban Redevelopment (NCPO) in 2008 to share their experiences and form strategies to improve their situation. Since the creation of the NCPO, more organisations have been established as part of the same movement.

Owner-occupiers opposed urban redevelopment and formed movements with the slogan 'No urban redevelopment, Leave my home'. Owner-occupiers highlighted the token involvement of residents over urban redevelopment. They emphasised individual and collective self-determination in urban redevelopment processes and initiatives. After recognising the uneven balance of power, owner-occupiers activism demanded more citizen participation in order to challenge the injustice resulting from the actions of pro-urban redevelopment coalitions. This activism turned urban redevelopment projects around and changed the state's approach to urban redevelopment. The state was reluctant to directly address owner-occupiers' complaints, but owner-occupiers activism made the state more responsive to their needs. In particular, the Seoul Metropolitan Government (the SMG) has been supportive of owner-occupiers' opposition movements after a progressive cross-bencher who used to work at an NGO took power in the 2011 by-election for the mayor of Seoul. The current mayor of Seoul has tried to mitigate some of the problems caused by urban redevelopment. The most representative example of this is the Community Building Project in Seoul, which focuses on preservation and small-scale regeneration led by residents. This marks a turning point in urban redevelopment policy.

Owner-occupiers activism is responsible for the introduction of various new systems. It contributed to the changing role of the local state and to the cooperation between residents and the state in managing urban redevelopment. It succeeded in encouraging residents to become more involved in urban redevelopment areas and in the management process of urban redevelopment in their neighbourhoods. That is, this activism tried to get power over and control of urban redevelopment. In terms of the typology established by Pickvance (1985), owner-occupiers' opposition movements moved to Type 3. Owner-occupiers requested their right to participate in the decision-making process in order to limit the influence of speculative investors.

The Third Wave of Housing Rights Activism

These movements, from the first wave through to the second wave, are mainly self-help movements and have various limitations that keep them from developing into a larger framework that can bring about fundamental change. Tenants' and owner-occupiers' activism has made progress in developing strategies to address the adverse effects of market-driven urban redevelopment and has led to policy changes and new legislation. Tenants and owner-occupiers have tried to improve laws and policies to strengthen all citizens' housing rights. However, their activism has not seen the creation of new alternatives to the existing society—alternatives that could create better and stronger communities capable of standing against market-driven redevelopment.

In contrast, since 2012, when the mayor of Seoul introduced the Community Building Project, new movements within civil society have emerged in order to build an alternative urban redevelopment system and change the conventional concepts of housing and urban redevelopment. The SMG has been focusing on creating a sense of community and growing communities in urban neighbourhoods. This new urban redevelopment project has been introduced under the slogan 'From urban redevelopment for profit to urban redevelopment for dwellers'. Even though the state has attempted to solve various problems, there are still gaps between what people want and what the state does. Therefore, new forms of grass-roots activism have emerged. Urban regeneration with social- or community-interest enterprises (Lee 2012) and housing cooperative movements with community land trust organisations are some of the new movements to have emerged. Also, new social organisations have been established to cover more specific groups' needs. Minsnail Union, which has tried to protect youths' housing rights, is one of them (Lee 2013a). These alternative movements aim to lead to the provision of affordable housing with secure tenancy and decent jobs to low-income households through self-sustaining programmes. The union has tried to improve laws and policies to strengthen all citizens' housing rights. These new movements have gained a degree of input with regard to housing development decision-making, which just a few powerful groups used to control. They have insisted on taking on the role of the state in local community services and provided infrastructures, which the state has failed to do. This new phenomenon has generated pressure for a new approach to be taken. Despite the lack of collective identities within the groups and larger goals for the public, these new movements have placed a limit on the old system and made the state turn its attention to innovative policies that accommodate social justice.

These new attempts could offer a fresh approach and a social alternative to the prevalent bureaucratic, top-down-managed and capital-driven systems. This new form of housing consumption and provision through housing cooperatives and new urban redevelopment methods could be one strategy that makes it possible to realise consumer control over housing on the basis of need rather than producers' interests and speculator profiteering. These movements also offer tenants an alternative housing tenure system instead of the current unfair system. Accordingly, people can have control of their homes rather than having them being controlled by the state. Tenants and some owner-occupiers, who have hardly been involved in existing housing and urban redevelopment policies, can become empowered to collectively control their neighbourhoods, and their communities can be rebuilt in a cooperative way. This urban activism is growing more successful.

The third wave of movements is still in its early stages, but these movements are valuable starting points that make it possible to think about what kind of city we want to live in and how to reshape our cities by ourselves. These grass-roots movements potentially stand to create fundamentally different cities and change people's day-to-day lives, since each movement has led to general questions in society about inequality and unfairness. The answer to these issues is a just society that guarantees distributive justice and makes greater socio-spatial equality a reality. As a result, new topics such as economic democracy, universal welfare and housing rights are receiving greater attention, and discussing them stands to help create a more equitable, just and democratic society for the 'have nots': a better society in socio-economic, political and spatial terms. This new grass-roots approach could be effective at tackling wider structural inequality that has resulted in unequal resource allocation for economic growth over the last half-century. It is too early to conclude that all these new movements will bring about fundamental changes to improve democracy, equity and diversity. However, they have drawn the attention of the state and society to creating a just city, rather than prioritising urban growth, and have offered the possibility of establishing an alternative urban development system.

Conclusion

There has been a lack of research into the issues of civil society in East Asian developmental states (Pekkanen 2004, p. 363). This chapter has raised important questions about the evolution of civil society in Korea. When looking at urban activism over housing rights in Korea, the institutional legacy of the developmental state needs to be considered. Social protests over housing rights can be understood in terms of the emergence and development of urban activism for fundamental social changes. This understanding can be developed by examining the relationship between the transformation that broad urban activism has brought about and the political environment. In spite of the progress made in the political environment, the classic instruments of citizen participation are still limited when it comes to urban restructuring and housing. In order to increase opportunities for citizen participation, it is crucial to think about a new way of overcoming the legacy of the developmental state. As Fainstein (2010, p. 181) underlines the importance of citizens' activism when it comes to the development of just policies, everyday democracy and active citizenship are suggested as ways to improve individuals' participatory rights in local politics.

This chapter has examined the evolution of housing activism in Korea over the last four decades in terms of the evolution of housing rights activism and has explored new alternative movements by civil society that challenge the dominant social power relations in a (post-)developmental state. In short, Pickvance's Type 4 defensive movements against threats induced by urban redevelopment have prevailed in Korea, since Korean cities have developed very rapidly. However, the goals of housing rights activism have led to and overlapped with the other types of movements over time. Many Koreans' rights and entitlement to be involved in housing issues have been inadequate. It has been very difficult to bring about institutional transformation by collective action. Social struggles have not succeeded in changing the dominant discourses of housing development. However, the analyses of opposition movements led by owner-occupiers and alternative movements imply a paradigm shift in housing rights movements in Korea to a certain extent. The analyses suggest that citizens are conscious of their participation in controlling urban redevelopment processes. These new movements have promoted alternatives that are more democratic and just forms of housing development. New attempts at the grass-roots level have been made in order to minimise the various problems caused by the powerful alliance between the state and the capital over housing and urban redevelopment. People are challenging the wider capitalist processes of urbanisation. A variety of reforms have been attempted, with consideration having been given to universal welfare and fairness. The current changes are in response to the crisis of the long-lasting housing policies that are the institutional legacy of the developmental state. Changes in Korea are in progress, having been pushed forward by growing pressure from citizens who have lost their trust in the old housing paradigm. There is no panacea for housing problems, since they are multifaceted and fluid. However, the new changes are signposts that will lead to better solutions.

References

Castells, M. (1976). Theoretical Propositions for an Experimental Study of Urban Social Movements. In C. Pickvance (Ed.), *Urban Sociology: Critical Essays*. Kent: Tavistock Publication.

Castells, M. (1983). *The City and the Grassroots*. Berkeley: University of California Press.

Cho, O. R. (1989). Urban Poor's Socio-Economic Characteristics and Community Movements. In H. K. Kim (Ed.), *Slums and Urban Redevelopment*. Seoul: Nanam (in Korean).

Davis, L. K. (2011). International Events and Mass Evictions: A Longer View. *International Journal of Urban and Regional Research, 35*, 582–599.

Doling, J. (1999). Housing Policies and the Little Tigers: How Do They Compare with Other Industrialised Countries? *Housing Studies, 14*, 229–250.

Fainstein, S. S. (2010). *The Just City*. New York: Cornell University Press.

Fainstein, S. S., & Fainstein, N. I. (1985). Economic Restructuring and the Rise of Urban Social Movements. *Urban Affairs Review, 21*, 187–206.

Ha, S.-K. (2002). The Urban Poor, Rental Accommodations, and Housing Policy in Korea. *Cities, 19*, 195–203.

He, S. (2012). Two Waves of Gentrification and Emerging Rights Issues in Guangzhou, China. *Environment and Planning A, 44*, 2817–2833.

Hwang, I. C., & Yoo, N. Y. (1989). Urban Poor's Legal Status Related to Residence: Urban Redevelopment and Tenants' Rights. In H. Kim (Ed.), *Slums and Urban Redevelopment*. Seoul: Nanam (in Korean).

Jang, K. S. (2006). The Spatial Development of Developmental State: The Case of the Process of Residential Area Formation Along the Han Riversides

During the 1960s and 1980s in Seoul, Korea. *Space and Society: Journal of Korean Association of Spatial and Environment Research, 25*, 194–212 (in Korean).

Kim, D.-C. (2006a). Growth and Crisis of the Korean Citizens' Movement. *Korea Journal*, (Summer), 99–128.

Kim, H. H. (1998). South Korea: Experiences of Eviction in Seoul. In A. Azuela, E. Duhau, & E. Ortiz (Eds.), *Evictions and the Right to Housing: Experience from Canada, Chile, The Dominican Republic, South Africa, and South Korea*. Ottawa: International Development Research Centre.

Kim, H.-R. (2003). Unraveling Civil Society in South Korea: Old Discourses and New Visions. In D. C. A. W. H. Schak (Ed.), *Civil Society in Asia*. Aldershot: Ashgate.

Kim, J. (2010). *Mobilizing Property-Based Interests: Politics of Policy-Driven Gentrification in Seoul, Korea*. PhD, University of Illinois.

Kim, K. H., & Cho, M. (2010). Structural Changes, Housing Price Dynamics and Housing Affordability in Korea. *Housing Studies, 25*, 839–856.

Kim, S. H., Chun, H.-K., & Jang, J.-H. (1998). *Eviction from the View of the Evicted*. Seoul: Korea Centre for City and Environment Research.

Kim, Y.-M. (2006b). *Towards a Comprehensive Welfare State in South Korea: Institutional Features, New Socio-Economic and Political Pressures, and the Possibility of the Welfare State*. London: Asia Research Centres.

Kwon, S., & Holliday, I. (2007). The Korean Welfare State: A Paradox of Expansion in an Era of Globalisation and Economic Crisis. *International Journal of Social Welfare, 16*, 242–248.

Leckie, S. (1992). *From Housing Needs to Housing Rights: An Analysis of the Right to Adequate Housing Under International Human Rights Law*. London: International Institute for Environment and Development.

Lee, H. K. (1999). Globalization and the Emerging Welfare State-the Experience of South Korea. *International Journal of Social Welfare, 8*, 23–37.

Lee, H. S. (2013a). Youth's Housing Rights, Minsnail Union. *Urbanity and Poverty, 102*, 49–61.

Lee, J. (2003). Is There an East Asian Housing Culture? Contrasting Housing Systems of Hong Kong, Singapore, Taiwan and South Korea. *The Journal of Comparative Asian Development, 2*, 3–19.

Lee, J. W. (2012). Building Community, Housing Regeneration by Social Enterprise: A Case Study of Toadhousing. *LHI Archives, 9*, 106–113.

Lee, J. Y. (1990). *The Practice of Protest: Three Case Studies in Urban Renewal in Seoul, Korea*. PhD, City University of New York.

Lee, J. Y. (2000). *The Practice of Urban Renewal in Seoul, Korea: Mode, Governance, and Sustainability*. The 2nd International Critical Geography Conference, Taegu, Korea (in Korean).

Lee, S.-H. (2013b). Counterattack of Apartments. Seoul: KBS (in Korean).

Lee, S. Y. (2014). *New-Build Gentrification and Anti-Gentrification Movements in Seoul, South Korea*. PhD, King's College London.

Lee, S. Y. (2016). Neil Smith, Gentrification and Korea. *Space and Society: Journal of Korean Association of Spatial and Environment Research, 26*, 209–234 (in Korean).

Lim, S.-H. (1995). Housing Environment in the Era of Local Autonomy. *Space and Society: Journal of Korean Association of Spatial and Environment Research, 6*, 102–124 (in Korean).

Lim, S.-H. (2005). *Housing Policies Over Half of the Last Century*. Seoul: Kimundang (in Korean).

Mayer, M. (2006). Manuel Castells' the City and the Grassroots'. *International Journal of Urban and Regional Research, 30*, 202–206.

McGovern, S. J. (1997). Cultural Hegemony as an Impediments to Urban Protest Movement: Grassroots Activism and Downtown Development in Washington, DC. *Journal of Urban Affairs, 19*, 419–443.

Ng, M. K. (2002). Property-Led Urban Renewal in Hong Kong: Any Place for the Community? *Sustainable Development, 10*, 140–146.

Park, Y.-N. (1988). Housing Policy and the Urban Social Movement. *Regional Study: Journal of Korean Regional Science Association, 4*, 59–71 (in Korean).

Park, I. K., & Lee, S. Y. (2012). Analyzing Changes in Space and Movements for Opposition and Alternatives in Seoul. *Space and Society: Journal of Korean Association of Spatial and Environment Research, 22*, 5–50 (in Korean).

Pekkanen, R. (2004). After the Developmental State: Civil Society in Japan. *Journal of East Asian Studies, 4*, 363–388.

Pickvance, C. (1985). The Rise and Fall of Urban Movements and the Role of Comparative Analysis. *Environment and Planning D: Society and Space, 3*, 31–53.

Rabrenovic, G. (2008). Urban Social Movements. In J. S. Davies & D. L. Imbroscio (Eds.), *Theories of Urban Politics*. London: Sage.

Ronald, R., & Jin, M.-Y. (2010). Homeownership in South Korea: Examining Sector Underdevelopment. *Urban Studies, 47*, 2367–2388.

Shao, Q. (2013). *Shanghai Gone: Domicide and Defiance in a Chinese Megacity*. New York and Oxford: Rowman & Littlefield Publishers.

Shin, H. B. (2008). *Driven to Swim with the Tide? Urban Redevelopment and Community Participation in China* (CASE Discussion Paper Series). London: Centre for Analysis of Social Exclusion, London School of Economics.

Shin, H. B. (2009). Life in the Shadow of Mega-Events: Beijing Summer Olympia and Its Impact on Housing. *Geoforum, 40*, 906–917.

Shin, H. B. (2013). The Right to the City and Critical Reflections on China's Property Rights Activism. *Antipode, 45*, 1167–1189.

Shin, J. S. (2010). *The 'Housing Class' Differenciation of Urban (Re)development in South Korea—Based on a Case Study of G 'New Town' Project and Home Owner's Movement in Northern Seoul*. Master, Yonsei University (in Korean).

Shin, J.-W. (2011). The Uniqueness of Korean Residential Capitalism in International Comparison. *Trend and Perspective, 81*, 112–155 (in Korean).

Smart, A., & Lam, K. (2009). Urban Conflicts and the Policy Learning Process in Hong Kong: Urban Conflict and Policy Change in the 1950s and After 1997. *Journal of Asian Public Policy, 2*, 190–208.

Tarrow, S. G. (2011). *Power in Movement: Social Movements and Contentious Politics*. Cambridge: Cambridge University Press.

Weinstein, L., & Ren, X. (2009). The Changing Right to the City: Urban Renewal and Housing Rights in Globalizing Shanghai and Mumbai. *City & Community, 8*, 407–432.

Woo, M. S. (2004). Explaining Early Welfare Policies in South Korea: Focusing on the Nexus Between the State and the Business Sector. *Development and Society, 33*, 185–206.

Wu, F. (2004). Residential Relocation Under Market-Oriented Redevelopment: The Process and Outcomes in Urban China. *Geoforum, 35*, 453–470.

Yoon, I.-S. (1994). *Housing in a Newly Industrialized Economy: The Case of South Korea*. Aldershot: Avebury.

CHAPTER 12

The Squatters' and Tenants' Movement in Buenos Aires. A Vindication of Centrality and the Self-Managed Production of Space

Ibán Díaz-Parra

Introduction

In the 1990s, the local government of Buenos Aires renewed its interest in revitalising its city centre, as many other Latin-American cities were doing at the time. Massive private investments in urban renewal lead to the conversion of old and declined neighbourhoods and the creation of new artistic districts intended to be part of the contemporary urban vision for the city. However, these projects faced significant obstacles as a result of national economic and political crises as well as the reluctance of the working class to abandon the central areas. In that regard, the Squatters and Tenants' Movement (MOI, in its Spanish acronym) has played a significant role in the recent evolution of Buenos Aires's city centre. It has helped transform squatted and ruined buildings into new neighbourhoods through cooperation and the self-management of collective property, creating a notable network of residential and social spaces.

I. Díaz-Parra (✉)
University of Seville, Seville, Spain

N. M. Yip et al. (eds.),
Contested Cities and Urban Activism, The Contemporary City,
https://doi.org/10.1007/978-981-13-1730-9_12

In addition, MOI demonstrates an explicit validation of the right to the city as well as referencing many other Lefebvrian concepts in its public documents and militant speeches. These aspects make MOI a movement with a strong spatial focus in its practices as well as in its discourses.

Today, MOI is a strong example of the historical vindication of the right to centrality by the working class in Buenos Aires and demonstrates the power of popular organisations to create and recreate an urban space. This study contributes to growing research on spatiality in grassroots movements (Nicholls et al. 2013), and it focusses on Lefebvrian analytical concepts to do so. In recent years, some authors have studied the capacity of grassroots movements to define and create their own territory and to develop alternative spatial projects, often located in peripheral and non-urbanised areas (Mançano 2011; Agnew and Ostellender 2013). The capacity of grassroots and self-managed organisations to create spaces in central areas which are held by the state and private capital is debatable, as is the extent of impact that they have on the transformation and evolution of the internal structure of major cities. In this regard, the current article examines the capacity, potential and limits of contemporary urban activism and grassroots movements to produce autonomous spatial projects in densified urban areas, where most of the Latin-American population lives today. This chapter examines two aspects related to this issue. First, it questions the degree of independence which these grassroots movements can have from the state with regard to the production of space. Second, it probes the ideological dimension of the struggle for urban centrality, investigating the relevance of critical urban studies on the concept of ideology itself.

This first question is relevant because many authors may be overestimating the degree of autonomy which grassroots movements can have from the state. Popular management of the habitat is common in Latin-American cities, although it is limited to areas that have been partially abandoned by the state and/or private capital, usually in peripheral sectors. However, there has been a wide celebration of the capacity of the working class in Latin-America to produce their own habitat from both neoliberal (De Soto 2001) and libertarian discourses (Turner and Fichter 1972). More recently, from a far-left point of view, the self-managed production of habitat could be described as a non-alienated way of producing space (Zibechi 2007; Holloway 2010). This perspective can lead to overly optimistic assumptions about the capabilities of urban activism, ignoring the role of the state or reducing it to a simple antagonist of transformative politics. Other discourses on the social production of habitat (SPH), including that of MOI, do not espouse a romantic discourse on self-management; they point to the problems resulting from the fragmentation of autonomous grassroots organisations and to certain state institutions as a battleground of struggle and dispute (Rodríguez 2009).

With regard to the question of ideology, through its struggle for central places, MOI also confronts capitalist ideological patterns: accumulation for accumulation's sake, the logic of maximum profitability and the ideologies of private property and economically rational decision-making. MOI's struggle is for physical space but it also conducts an ideological battle for the legitimacy of alternative ways of occupying space. Based on the principles of autonomy and self-management, MOI has developed a movement for occupying the scarce land of Buenos Aires's central areas, competing with more profitable land uses, channelling state economic political support as well as organising public protests and collective squatting. This has required the recognition of collective living rights and suggested an alternative vision of what the city is and who has a legitimate right to live in it. As a result, MOI's housing cooperatives face the unavoidable conflicts that arise from friction between consumer individualism, and a vision of collective property and shared living.

The research presented here is based on 11 qualitative interviews with MOI cooperative members. It is focussed on their personal, residential and political trajectories, combined with a period of three months of participant observation inside the organisation's work areas and housing cooperatives. This fieldwork also involved the analysis of documents published by the organisation and an exhaustive study of secondary sources regarding recent urban development and planning strategies in the city of Buenos Aires. As a result of the research, this chapter advances two main conclusions. First, in densified and consolidated urban areas, under the present conditions, grassroots movements cannot produce new spatial orders and autonomous practices but, rather, must always negotiate with the state's own socio-spatial interventions. Second, examining the term ideology is important in keeping urban studies politicised. The concept of ideology raises the question of the possibility or impossibility of transforming socio-spatial orders and therefore addresses a key political question.

Lefebvre and the Right to the Centrality

Throughout the conversations with housing cooperative members during the fieldwork, Lefebvre was very present, even if nobody mentioned his name. References to the right to the city, habitat self-management or the priority of use value over exchange value in the urban space were very common in the interviews with those who were not academics. As a critical analyst using a Lefebvrian analytical framework, I discovered that the majority of my interviewees shared a view of their own reality which supported Lefebvrian concepts. This is a highly uncommon situation in ethnography, where there is often a significant gap between the etic perspective of the researcher and the emic practices of the group that the researcher is studying. This overlap is the result of the alignment of key organisation members with the ideas of the French philosopher. Maria Carla Rodriguez, the academic who has conducted the majority of the research into the history of this organisation, is at the same time one of its key members. Thus, MOI is an optimal example of the interrelation between critical academy and grassroots activism.

Several authors have previously written about MOI. Diaz Orueta et al. (2001) focussed on MOI within the context of the urban restructuring of Buenos Aires in the nineties, and Zapata (2016) studied the relation of social grassroots organisations with local state housing policies in Buenos Aires. Maria Carla Rodriguez (2004) has been the principal biographer of the organisation, researching its politics and its organisational culture (2009), mainly from the framework of the SPH. The SPH conceptualisation of habitat production was developed in Latin America in the nineties for describing non-mercantile ways of producing habitat, based on a logic of necessity and non-profit, as part of the production of illegal settlements by lower classes and for the cooperative and collective production of housing and neighbourhoods (Rodríguez et al. 2015). Although SPH is not a Lefebvrian concept, it shares some parallels with regard to the production of space as determined by the logic of different modes of production (Lefebvre 1992). Orueta, Zapata and Rodriguez each make some use of Lefebvre's work, especially the currently ubiquitous concept of the right to the city. The latter is used (especially by Rodriguez and Orueta) with regard to self-management and participation in the construction of the urban habitat, that is, as the right to create and transform the city. Apart from these aspects, most of Lefebvre's remaining conceptual arsenal has not been utilised in studying MOI. Most of the analyses have occurred on a more concrete level, studying the effects of neoliberalism, structural adjustment and the deterioration of living conditions on housing. With this in mind, this chapter contributes to three fundamental areas by analysing MOI'S spatiality through the filter of Lefebvre's work: (1) examining 'the right to the city' as a political claim that focusses on the struggle for centrality; (2) considering urbanism as an ideology; and (3) evaluating the limits of an autonomous, non-market, non-state agent for producing space in urban centres.

Although self-management is very present in the right to the city claim, in Lefebvre's work, it also has everything to do with centrality. For Lefebvre, centrality arises intrinsically with the emergence of the city, as both are the result of the historical concentration of wealth and power (Lefebvre 1969, p. 19). Hence, centre-periphery makes up the most substantial spatial dimension of the city and is imbued with a strong class content. In his book *The Right to the City*, the city is a synonym of city centre, and the eviction of the working class from central areas to new and peripheral neighbourhoods (exemplified in Haussmann's urban reform of Paris) is equivalent to an expulsion from the city itself. For low classes, the right to the city is the recovery of the city as well as the recovery of central areas, not just a right to 'visit or occasionally return' (Lefebvre 1969, p. 138). On the other hand, working-class expulsion from the city centre leads to the latter becoming a 'high quality product for the consumption of foreigners, tourists, people coming from peripheral neighborhoods, suburban commuters; surviving thanks to a double function: place for consumption and consumption of place' (Lefebvre 1969, pp. 27–28). This tendency has been reinforced in recent decades in Latin-American cities, where the utilisation of historical heritage for the attraction of tourists and local visitors is playing a key role in the renewal of central areas (Diaz-Parra 2014).

Moreover, changes in contemporary capitalism have led to an intensification of the role of urban centrality for the economic functions of command and control, in parallel to providing exclusive areas for urban professional elites. The city is a pole which centralises resources resulting in an unequal distribution. As wealth and prestige centralise, this simultaneously produces peripheries. Around the city centre (or centres), the land turns scarce and acquires an exchange value, thus becoming exploitable as a commodity for landlords. Every piece of land has an exchange value in the contemporary city, but the scarcity and the exchange value of the land itself is created by the production of centrality (Lefebvre 1992). The monopolisation of land as a commodity divides space into multiple properties and its differential value creates a geography of socio-spatial segregation in the city. 'That is how space turns into the means of segregation and the dispersion of social elements expelled to the peripheries' (Lefebvre 1992, p. 368). In this sense, the right to the city is a claim to recover the 'appropriated' city and an oppositional demand that challenges the rich and powerful and calls for a redistribution of wealth, means of action, space, etc. for those deprived (Mayer 2012, p. 71).

This struggle directly confronts urbanism as an ideology serving those in power. Ideology, in its critical Marxist sense, consists of an 'inversion' in which capitalist contradictions become their opposites. Capitalism's site of 'appearances' is, of course, the market, where injustice and inequality appear under the guise of justice and equality, naturalising and legitimising capitalist relations where human beings become alienated from the objects that they create (Larraín 2007). Lefebvre denounces urbanism (at least in his first books on the urban) as a superstructure which naturalises and legitimises capitalism and allows for organising the social reproduction of mass consumerism. Urbanism creates habitats oriented to different consumption paths; it makes sacred the moral function of private property and conforms individuals to their place in the social hierarchy, at work, as well as in their neighbourhoods (Lefebvre 1969, 2003). The consumption of space is the consumption of lifestyles, where neighbourhoods become spaces of representation with precise meanings in a symbolic system shared by society. Thus, the struggle for space and the production of space are not simply physical issues, nor are they merely conflicts relating to the concentration and distribution of material resources. They also entail the production of knowledge and meanings, ideological readings of space for supporting social reproduction according to capitalist demands, and the production of particular behaviours and lifestyles (Lefebvre 2013).

In the end, urban space functions as a mediator between the paths imposed by global capitalism and the daily life of individuals, but it might also provide a means for individuals to, in turn, act upon structures. As space is the unavoidable mediator of all social practices, Lefebvre speaks of spatial practices (projecting onto a spatial field all aspects, elements and moments of social practice) and spatialised social relations as always being subjected to political practices, to control and manipulation by the state (Lefebvre 1992). Other authors consider the spatial mediation of social practices as potentially serving different political objectives, or the particular interests of specific classes, groups or individuals (León et al. 2009). A dispute between antagonistic spatial projects necessarily entails an ideological struggle. Therefore, one could elaborate that for Lefebvre, the production of space in capitalist ideology is an alienated way of producing space, in which workers are deprived of their production (of space) and enslaved by their products: houses, habitats and land markets, condemned to peripheral locations and precarious habitats. In this light, the very idea of SPH coincides with the project of a non-alienated way of producing the city. However, not all self-productions of habitat by lower classes are SPH. Only those in which the collective retains control over the entire production process as well as the finished product are considered SPH (Rodríguez et al. 2015).

Back to the City in Buenos Aires

From the nineties onwards, a shift in urban politics towards a neoliberal ideology occurred in Argentina as well as in most of Latin America. Carlos Menem's government promoted the liberalisation of the economy and the privatisation of public enterprises, which coincided with an increase in the flow of foreign capital. As a result, a main trend in the urban politics of Buenos Aires was recentralising, densifying and reinvesting in the historical central district (Ciccolella 1999). In the last two decades, local governments have promoted urban renewal in the southern sector of the city centre, which has perpetuated the radical segregation of many downgraded and low-class neighbourhoods from the affluent northern sector of the city (Herzer 2008). Urban political strategies have focussed investments on developing under-utilised industrial or transport-based infrastructures and declined residential areas. This type of urbanism has emphasised place-based projects oriented towards the organisation of a consumption-based lifestyle. The strategy it used was one of traditional physical urban renewal but with heightened attention paid to the symbolic valorisation of place. Instead of appealing to consumers through high-quality developments, there has been a growing importance given to ideological investment. These developments demand a production of diverse spatialised meanings (ideological organisation of consumption) such as ostentatious consumption as well as the creation of folkloric and patrimonial places and artistic neighbourhoods.

In spite of their diversity, all of these projects have in common that they target solvent consumers: traditional urban bourgeoisie, foreign tourists, young academics and cultural entrepreneurs. Projects demonstrating these features were initiated in the nineties and continue to transform the central area of Buenos Aires today.

Puerto Madero was the flagship of urban renewal in Buenos Aires in the nineties, involving the privatisation and redevelopment of the harbour. As a result, a new urban sector oriented towards transnational capital and high-quality consumption have emerged: offices, restaurants, luxury hotels, yacht marinas and high-income housing. Rodríguez et al. (2008) inferred a neoliberal ideological perspective into this development because the public investment was directed to promote private profits. In addition, Ramirez (2011) points out how the presence of European influences, such as the model of London Docks and an agreement with Barcelona's local state for developing the strategic plan, has resulted in the creation of a globalised aesthetic product.

Puerto Madero was the first step in an urban renewal project that continued with the redevelopment of the harbour at La Boca. This operation had an extensive impact on this low-class neighbourhood, requiring the complete redevelopment of a clearly delimited sector (Vuelta de Rocha and Caminito) with shops and restaurants aimed at tourists and local visitors. Today, the result is a touristic area that is highly thematised around the culture of Tango, where old, substandard houses have been transformed into colourful souvenir shops.

Another key piece of the urban restructuring of Buenos Aires was the renewal of the historical city centre, bounded by the neighbourhood of San Telmo. The strategic plan for the historical city centre was intended to promote cultural tourism and industry, while improving public space and restoring highly deteriorated housing, mostly composed of collective rented housing for lower classes. Rodriguez et al. (2008, p. 84) see in the project the aim to create a place trademark for San Telmo. As a result, there was a strong revalorisation of the area in the nineties, stopped by the crisis in 2000 and with a new boom after 2004 (Ostuni 2008, p. 246). Today, San Telmo is a 'must-see' area for visitors to Buenos Aires. Many of the old collective rented houses have been turned in hostels for foreign backpackers.

Another relevant urban renewal project initiated in the 1990s is the El Abasto area, which lies outside the southern corridor, in the commercial city centre. El Abasto had always been a prestigious area, named after the impressive Art Deco market building. It was considered to be one of the main areas of tango culture and includes the home where Carlos Gardel lived. However, at the end of the eighties, El Abasto had become stigmatised due to the presence of numerous precarious collectively rented houses and squatted buildings. In the nineties, the El Abasto market building was bought by a company owned by George Soros, who transformed it into a vast commercial space, coupled with three high-income apartment towers. As a consequence, most of the squatters in the area were evicted, usually without violence, in exchange for money. Since 1999, there has been an intensive heritage-based valorisation in the area, with the creation of new thematic restaurants, hotels, antique shops, theatres and museums. Carman (2006) highlights the removal of undesirable neighbours and the gentrification of space as a result of this urban renewal.

Finally, in the nineties, declining neighbourhoods in Old Palermo and Pacific Palermo, also in the city centre, were renamed as Palermo Soho and Palermo Hollywood in a city branding strategy. These areas were formerly occupied by small industries and shops. Today, they contain dozens of highly specialised shops and restaurants, displaying typical artistic and leisure—district aesthetics (Carbajal 2003). This tendency has been reinforced by the creation of a district business by the local government, aimed at encouraging the settlement of entrepreneurs and companies in the area.

Altogether, there has been a high revalorisation of these areas, especially in the southern part of the city centre. This urban renewal is also an ideological reorganisation of the internal space of the city where centre-periphery logic is reinforced by investment in central areas while expelling the working class to under-invested, under-equipped and stigmatised peripheral neighbourhoods.

Struggling for Land in the City Centre

Access to centrality in Buenos Aires, as in many other large cities, has a special meaning. Jobs, public administration and public services are highly centralised in a city of 15 million inhabitants (Greater Buenos Aires), while the public transport system remains limited outside the central area. Heavy traffic and gridlock force many residents to spend two or more hours commuting each day. Therefore, this is a case in which the right to the city is equivalent to access to centrality since those living in peripheral working-class areas are deprived of access to the place where social products are concentrated.

The vindication of centrality is a key issue in Buenos Aires and it is also an old one. During the military dictatorship, urban renewal and highway development were undertaken to strategically erase marginal settlements from the inner city. In addition, elimination of rent controls allowed for a mass eviction of tenants from the centre. The restoration of liberal democracy in the late 1980s and early 1990s resulted in the proliferation of squatting in abandoned city centre buildings, within an environment of public tolerance and reduced budgets. This was a disobedient spatial practice which undermined the logics of segregation and resisted the expelling of undesirable social elements towards peripheral areas, through the occupation of interstitial and abandoned spaces in the city. Even if the growth of precarious and illegal settlements had its peak previously, in the 1980s there was a rapid increase in the population of the existing illegal settlements in the central areas. This was accompanied by the occupation of new land in old industrial and infrastructural areas, as these were the only options for the new immigrant and poor populations to access centrality. The squatting of old empty and deteriorated buildings was a new and growing phenomenon in the 1980s (Rodríguez et al. 2015). Much of this squatting of empty buildings was organised by families evicted from rented houses and precarious hotels in the city centre (Carman 2006).

The end of the dictatorship also allowed for the return of repressed left-wing academics (mainly architects), who had been barred from the university and were now looking to involve the academy in social struggles. 'This confluence allowed the creation of a social organisation oriented towards the struggle for the right to the city' (MOI 2012, p. 71). These academics began to work with some of the squatting groups in the city centre, promoting self-organisation of families. Later, this group would be influenced by the Uruguayan Federation of Mutual-Aid Housing Cooperatives (FUCVAM in Spanish) and would attempt to replicate its model.

In 1987, these activists connected with the squatters of a state-owned building in San Telmo which housed dozens of families. They developed a proposal of restoration and self-management of the space through a collective-owned and non-profit cooperative housing scheme, reclaiming both the families' right to live and to remain in the city centre. The process was curtailed as the result of a change in government in 1992 and the dwellers were evicted in 2003, after a strong defamation campaign (Rodríguez 2009). In fact, during the 1990s there was a shift from tolerance to hostility from the state towards squatters, including 'exemplary evictions' and a stigmatisation campaign with a racist discourse against immigrants (a significant proportion of squatters and people housed in precarious conditions in Buenos Aires were and are foreign immigrants). Generally, the evictions were linked to the tightening of immigration laws (Carman 2006, pp. 64–65).

At this time, new connections were being forged between the activists and other squatter groups in the city centre, as a result of their contact with the families in San Telmo. Through systematic meetings and daily interactions with the families, the activists advocated for collective work in these squats, including the improvement of their internal organisation, establishing channels for dialogue with local government and creating a cooperative project involving the institutionalisation of the assembly of squatters as a legal cooperative. The main reason current MOI members give for changing their situation from being squatters to becoming part of a legally instituted cooperative was to improve their living conditions and reach some stability for their families. Living conditions in the squats can be rough. The threat of eviction is permanent, and families are burdened with stigmatisation by their own neighbours while drug dealing and extreme marginalisation create situations of internal violence within the squat. These early groups of squatter families were the basis for the creation of a new organisation; a federation of housing cooperatives under the name of MOI in 1990. That year, MOI was integrated into the Latin-American Secretary of Popular Housing (SELVIP in Spanish) network. Furthermore, in 1994, MOI joined the Argentine Workers' Union (CTA in Spanish), as the habitat division of this leftist union, an alliance which remains in place today.

In many cases, the activists did not succeed in convincing a majority of families within a single squat to join a cooperative process. As a result, MOI began working with a range of people with housing problems and not only squatters. They created organisational spaces called Guardias where people could form or become part of a cooperative federated under MOI. In the early nineties, MOI created over 15 cooperatives, involving nearly 600 families, most of whom had participated in previous squats. Within the federation, each group was instituted as a formal cooperative with its own autonomous assembly. Each assembly followed MOI's basic criteria for the production of space: mutual aid, self-management and collective property. MOI also gave rise to notorious charismatics and informal leaders, who (in this case) seem to be crucial for the continuity of the federation as a cohesive movement with a shared political discourse.

Most of the projects collapsed without achieving their objectives for one reason or another. Nestor Jeifetz, one of the main leaders of MOI, recalled that 'from each of the three or four squats we were working with, we succeeded with one' (MOI 2012, p. 7). A few of the cooperatives—Yatay, La Unión, Fortaleza and Perú—accomplished their objective of buying plots of land in the city centre. La Unión, Fortaleza and Perú were originally squatted buildings; Peru in San Telmo, Fortaleza in the area of Puerto Madero and La Unión also in a location close to the historical city centre. Yatai was born of families who had come from evictions of other squatted buildings. The cooperative bought an empty plot of land in Barracas, a relatively central area, south of the historical city centre of Buenos Aires, and built a new housing development there. In addition to these initiatives, MOI created a project of self-managed and transitory housing, buying and restoring a building in 1996, also in Barracas, for the members of the cooperatives to dwell in while their future houses were being developed. Later, two more buildings were added to this system of transitory dwelling.

In 2001, the economic and political crisis in Argentina led to government instability and to the emergence of a new kind of social protest: neighbourhood assemblies, piqueteros, occupied factories, and so on. In this context, the government of Buenos Aires opened a space for debating the housing crisis with social organisations. As a result, Act 341 was created, a law which cooperative members call 'Our Act' and 'the only self-managed urban development act in the country', as it exists only in Buenos Aires. The law ensures government credits for the cooperatives and leaves the control of the design, internal organisation and management of resources in the hands of the coop members. It makes explicit the need to transform the social problem of housing, encouraging self-management and cooperative formulas. In addition, Act 341 allocates state-owned plots of land located in the city centre for the development of cooperative projects as well as for the MOI transitory dwelling programme. While MOI played an active part in drafting the Act, it is today only one among many organisations who utilise it. MOI's main differentiating factors in comparison with other housing organisations are its commitment to the collective property and the use of mutual aid in the development process.

With the passing of the new act, MOI developed its two most ambitious housing cooperative projects to date: La Fabrica and El Molino. La Fabrica is a plot of land of 2500 m^2 in Barracas. It is a project made up of 50 duplexes from 60 to 90 m^2, with two multipurpose rooms, shops, squares and green areas. El Molino, near Parque Patricios, is a project composed of 100 duplexes with a three-floor community social centre, which currently accommodates a free nursery and a night-time secondary school (in both cases, workers are paid by national state funds). Co-op members' profiles are similar in both cooperatives: working-class families, coming from collective rental houses or squats, who lived for various years in the transitional housing programme. They are a culturally diverse group—many coming from the north of Argentina as well as from other countries including Chile, Paraguay, Uruguay, Bolivia and Peru. Both La Fabrica and El Molino share the neighbourhood with private luxury condos and middle-class housing, as a result of the revalorisation of these areas. As co-op members explained, these areas are currently a 'territory in dispute'. Another relevant issue in the cases of La Fabrica and El Molino, as well as in the case of Yatai, was the initial opposition of some of the neighbours to these settlements. As a Yatai cooperative member said 'Every place where the working class settle they find opposition'. Eventually, these early conflicts were resolved. In this way, the co-op's community services are aimed at improving not only the lives of the co-op members but also that of the whole neighbourhood, which facilitates integration.

Currently, MOI has six housing cooperatives in the city centre of Buenos Aires. All of them, as well as the buildings for transitory dwelling, are located in the south of the city centre. These six were created before 2003. In 2007, a conservative shift in the government of Buenos Aires resulted in the stagnation of development processes. MOI's projects are unavoidably affected by everyday politics, electoral politics as well as greater political events. Echeverria (1998) made a useful distinction between two types of politics, exceptional event politics and everyday politics. In this regard, the political exceptionality related to the 2001 crisis and the transition from dictatorship to democracy seem to be key events in the emergence and consolidation of MOI, whereas the organisation's involvement in everyday politics has been more problematic (Zapata 2012; Guevara et al. 2011). This point will be explored in more depth later in the chapter.

The Right to the City Versus the Right to Individual Property

MOI's spatial project clashes with the neoliberal project of internally restructuring Buenos Aires through the state and capital. The right of the working class to remain in the city centre opposes tendencies towards spatial reordering and segregation, as instruments of accumulation through the organisation of consumption and the production of space. In this regard, MOI offers an alternative representation of the city as well as an alternative proposal for daily life.

Recent neoliberal-oriented initiatives to restructure the city for consumption represent an ideology of spatial practice which legitimates some uses while condemning others. On the supply side, the best use is considered to be the most profitable one, following the logic of accumulation for accumulation's sake. Thus, private capitalists are legitimated in organising entire areas of the city in the pursuit of their own benefit, supposedly paying it back to the city by attracting investments and generating a spillover of wealth. On the demand side, legitimacy is established by one's social power converted into buying power. Hence, space is owned by the one who pays more, following the logic of private property as well as consumer rationality and freedom. All of this appears in political and technical discourses as the natural order of the city and is implicit in the urban renewal policies of the last two decades. Thus, central places are only considered efficiently used when desirable for solvent consumers. In a context of material and symbolic revalorisation of the city centre, the place for the working class is clearly the city's periphery. These elements are part of the current dominant urban ideology (see Table 12.1).

The concept of ideology is emphasised in Lefebvre's first works on the city with a negative and critical usage, as shown in the second epigraph. On the other hand, Larrain (2010) sees the positive usage (ideology as a particular set of political ideas and beliefs to which people adhere consciously) and negative usage of the term as incompatible, although it is certain that urban popular organisations do share alternative representations of the city that opposes the dominant ideology. This brings us back

Table 12.1 The right to the city as an ideological conflict

Dominant ideology	*Alternative-utopic discourses and practices*
Exchange value	Use value
Legal right	Moral right
Private property	Collective property
Individualism	Solidarity
Consumer	Member of cooperatives
Argentinian/immigrant	Latin-American
Market	Movement/class
Supply and demand	Self-management and mutual aid
Social power of money	Work capacity
Accumulation for accumulation's sake	Welfare state as redistributor
State as entrepreneur	

to an old problem in Marxism: is ideology merely an expression of class domination or is it the main battlefield of politics?

Cooperative members seeking to join MOI must undergo a long process of training, self-organisation and material contributions and abandonment is not infrequent. Being part of MOI is quite demanding and involves high levels of responsibility and discipline for co-op members. Co-op members contribute with their savings, their work in the construction of the buildings and through their participation in working groups and meetings. The co-op members are obligated to actively participate in the self-organisation of the collective through an involvement in assembly and work groups. All decisions are made in the assemblies and the cooperative members must take on responsibilities in the process of construction, distribution of the houses, organisation of common spaces, solidarity with other cooperatives and so on. General cooperative members contribute physical labour, while specialised work is done by skilled workers from MOI's work cooperative. The co-op members must register the property under the name of the cooperative, turning their legal right to private property into a right of use, which can be subrogated to children and grandchildren of members. Members are not permitted to sell the property for the purpose of extracting profit.

The Guardias are the main channels for entering MOI. The initial training for new co-op members is based on the history and principles of the organisation: self-management, mutual aid and collective property. This training aims to internalise MOI's discourse and interpellates the new members as Latin-American workers, inviting them to join a struggle for the land as use value. A central idea is that the city is built by the workers and not by money itself. Even the most expensive and luxurious buildings in Puerto Madero, or the restoration of complex infrastructure in El Abasto, were completed by manual workers, who usually live in extremely precarious houses and neighbourhoods. Many of them are non-white immigrants condemned to live in under-resourced settlements lacking urban infrastructure. Hence, the same people who build the city centre are deprived of it as well as from the lifestyles publicised by urban ideology. In this regard, MOI vindicates the right to the city, as 'for working class, expelled from the city centre to the periphery, dispossessed from the city, thus expropriated from the best result of its own work, this right has a very special meaning and scope' (Lefebvre 1969, p. 167). The co-op members interviewed agreed that behind the housing cooperatives lies the ideal that the working class is entitled to live in the city centre, in dignified houses and even in beautiful ones. The duplexes in El Molino, La Fábrica and Yatai are houses of an undeniably high standard with regard to materials used in construction, physical space and aesthetics. One co-op member said that some government technicians have criticised their use of expensive facing brick. It is as if common sense implied that 'beautiful houses are just for people in the north of the city'.

MOI's discourse interpellates co-op members as working class and Latin-Americans. National identity is as important as a class. In training sessions as well as in personal interviews, a frequent idea for explaining collective property is to make reference to this as part of the history of Latin America's native peoples. Constructing and consuming through mutual aid and self-management is another key element that is constantly restated. In Lefebvre's words, 'the right to oeuvre (to participant activity) and right to appropriation (very far from property rights) are embedded in the right to the city' (Lefebvre 1969, p. 159). However, self-management in MOI entails autonomous management of public resources. The idea repeated by co-op members in the interviews was that the state's resources are the peoples' resources. Thus, the state's credits are not gifts but, rather, ways of giving back resources. An MOI leaflet defined self-management as 'the exercise of our own capacity for managing resources in the benefit of the collective. It is taking part in the decision-making process about the destiny of public resources'.

Co-op members do not generally know which house will be theirs until the project is finished. This serves the purpose of ensuring full commitment to the collective work. In addition, the mutual aid system allows for a reduction in costs and co-op members value its capacity to strengthen social ties within the group. However, as time goes on, social ties tend to weaken. As the work progresses, the unqualified manpower of the co-op members becomes less necessary. When the families gain their homes, the group tends to split into those who have acquired a militant commitment (a minority) and those that 'enclose themselves in their homes'. In this regard, a few housing cooperatives have suffered relevant internal conflicts. Some co-ops have stagnated and some others have dissociated from MOI. There are ongoing disputes inside the cooperatives between supporters of MOI principles and adherents to more conventional ways of organising spatial practices. Key informants consulted say that the roots of the problem are issues pertaining to property and self-management. The high-quality housing projects are situated in valuable areas of the city centre and thus their price values have increased. Personal interests and strategies clash with the impossibility of selling the house since it is part of the housing cooperative's collective property. Some of the co-op members interviewed accuse others of espousing the autonomy of the housing cooperative in opposition to MOI with the objective of changing the land tenure agreement.

Because every cooperative is autonomous, conflicts are solved internally. Each cooperative has a formal structure, with elected directors who have a responsibility to develop the mandates of the assembly under its scrutiny and for a limited period. This is how MOI cooperatives manage the problem of collective organisation. It is the election of this directive group which seems to be the most common area of dispute when internal differences arise around property, self-management and so on. In the cooperatives with ongoing conflicts (such as Perú or El Molino), the members have split into separate groups with their own leaders. Some of these conflicts have resulted in the cooperative abandoning the federation. If a cooperative decides to abandon the basic principles of MOI (self-management, collective property and so on), it will exclude itself from the federation. Perú was in this situation during my fieldwork, although the interviewed cooperative members believed that the self-exclusion of this co-op from the federation was temporary. Thus, the door appears to be open for reincorporation, so long as the politics of the cooperative realign with those of the federation once again.

Disputes have also arisen between those supporting the Peronist administration governing the country from 2003 to 2015 and those supporting the left-wing political opposition. Furthermore, malicious interventions from different political powers trying to provoke internal fractures are not uncommon. The core of the dispute is often: collective property, solidarity and use value in opposition to purely economic rationality, individualism, fragmentation and exchange value (see Table 12.1).

At a reception event for a new generation of co-op members, Nestor Jeifetz, probably the main cadre of MOI, argued that 'collective organisation is the movement and not any individual housing cooperative or group'. MOI's achievements are the result of the collective and organised struggle to gain land and public resources. What has happened in the past is that 'the movement creates housing cooperatives and when they gain their objective (the housing), they just fuck off'. 'The struggle against the privatisation of urban land is at the heart of the whole organisation'. A core issue that is evident in this discourse is MOI's approach to daily urban life based on solidarity and universality, versus the individual egoism sanctioned as unavoidable by the dominant ideology.

Conclusion

The struggle for the right to the city is a struggle for land use as much as an ideological struggle. Urbanism as an ideology was one of the key conclusions of Lefebvre's work which has largely remained unexplored. This lack of exploration seems to be the logical result of the abandonment of the concept by the philosopher himself in his later work on the city. Lefebvre (1992) abandoned the negative and critical concept of ideology in favour of 'representation' in a context where post-structuralism generalised and systematised ideological criticism to every set of ideas and beliefs. If everything is ideology, the concept itself loses its utility. It is worth asking oneself along with Jameson (1992): if every code loses, who wins? It would be tempting to answer that in this context the dominant ideology wins, the ideology of the capitalist mode of production, the internalised ideology that does not say that it is ideology, which appears as a form of common sense (Zizek 2003), and which is less present in the field of discourse while dominating the field of social and spatial practices. It is important to disagree with Lefebvre's relinquishment of the concept of ideology and affirm with Zizek that the critical concept of ideology is still essential for keeping critical studies and urban studies politicised. Even if the negative concept of ideology is problematic, it remains a strong critical concept for urban and social movement studies and its abandonment or substitution by other concepts (habitus, representations, discourses and so on) seems to be an open door for depoliticisation. A retreat from ideology by critical scholars could help to naturalise the dominant spatial practices and erase its links with the endogenous logics of the capitalist mode of production. The theory of ideology therefore seems indispensable for an anti-capitalist questioning of current spatial orders.

On the other hand, its use presents some unresolved questions. To what extent is MOI's approach to the city an ideological one, even if it is not the dominant one? Do their spatial discourses and practices have the same effects of naturalising and masking contradictions as the dominant ideology does? As a provisional response it must be pointed out that, besides the fact that MOI is not controlling state apparatuses, MOI's approach follows objectives that are quite opposite to the dominant ideology. Dominant ideology is a mechanism for social reproduction while MOI's antagonistic discourse and practices try to threaten the status quo and offer a utopic alternative to current socio-spatial configurations. It could be said that the opposite of dominant ideology is not an alternative positive ideology, but the utopian itself. If dominant ideology is the element that prevents us from imagining a society different from that of capitalism, utopia is the negation of dominant ideology, the projection of an alternative socio-spatial order that doesn't exist but could exist. Hence, if the current urbanism is the current dominant ideology, SPH offers a utopian proposition that cannot be completely realised in a capitalist society, as MOI's daily problems and conflicts show.

Even if the capacity of one grassroots organisation is obviously limited, this study has shown how the working class struggle as a whole has been determinant in the actual internal configuration of the city of Buenos Aires. This is especially true in exceptional political times, such as the crises and protests of 2000 and 2001, and the transition from dictatorship to liberal democracy when the reconfiguration of spatial and social orders seemed possible. If, as is usually said in Argentina, neoliberal ideology and policy irrupted into the country with the military dictatorship and have continued until the present, these times of political exceptions have allowed, to some extent, the emergence of utopian spatial projects opposed to the dominant urbanistic ideology. If the current city centre of Buenos Aires is still socially diverse and governed by logics not only of the solvent consumer, it is a result of these exceptional political moments. On the other hand, the limits to the autonomous and utopic spatial practices of grassroots organisations are more apparent in daily politics. It seems obvious that the self-management of space by popular organisation is much less viable in high-density urban areas than in peripheral and low-urbanised areas.

In urban areas, the spaces for urban activism and their potentiality for utopian spatial practices and projects are unavoidably overdetermined by the previous and current organisation of the space by the state, by its interest as well as by its blind spots.

The struggles for the city centre as well as ideological struggles are fought not only in the streets but also in state institutions. The projects of organisations such as MOI, as well as the spontaneous actions of the working class, are not alien to the issue of state politics. The main achievement of MOI has been the creation of Act 341, and the 'giving back' of resources from the state to the people. Militant housing cooperatives, as well as the working class as a whole, cannot compete in the market for central locations in a city like Buenos Aires today. That is why state mediation is essential in addition to autonomous production and/or transformation of socio-spatial configurations by grassroots organisations. State resources are necessary in order to redistribute access in an urban context. Furthermore, effective action on scales larger than local ones is vital in order to gain the right to the city. In this light, MOI's current principle objective is the nationalisation and wide government support for Act 341.

Although grassroots activist organisations such as MOI are crucial in producing urban utopias, it can be seen that their impact ultimately remains limited. In the end, even if the problems of housing and habitat are essential, the right to the city in Buenos Aires is related to other unavoidable issues such as transport infrastructures, the dynamics of centralisation–decentralisation, the distribution of job opportunities and services, and more. Today, the state is still the only agent with the capacity to act on such problems on behalf of the disenfranchised.

References

Agnew, J., & Oslender, U. (2013). Overlapping Territorialities, Sovereignty in Dispute: Empirical Lessons from Latin America. In coordinado por Walter Nicholls, Byron Miller y Justin Beaumont (Eds.), *En Spaces of Contention. Spatialities and Social Movements* (pp. 121–140). Farnham: Ashgate.

Carbajal, R. (2003). Transformaciones socioeconómicas y urbanas en Palermo. *Revista Argentina de Sociología, 1*(1), 94–109.

Carman, M. (2006). *Las trampas de la cultura. Los intrusos y los nuevos usos del barrio de Gardel.* Buenos Aires: Paidos.

Ciccolella, P. (1999). Globalización y dualización en la Región Metropolitana de Buenos Aires. Grandes inversiones y reestructuración socioterritorial en los años noventa. *EURE, 15*(76), 5–27.

De Soto, H. (2001). *El misterio del capital: por qué el capitalismo triunfa en occidente y fracasa en e resto del mundo.* Lima: Sudamericana.

Díaz Orueta, F., Louré, M. L., & Agulle, J. M. (2001). *Ciudad, Democracia y Movimientos Sociales: el Movimiento de Ocupantes e Inquilinos de Buenos Aires.* VII Encuentro de latinoamericanistas españoles Madrid, 13–15 November.

Díaz-Parra, I. (2014). El regreso a la ciudad consolidada. Ciudades. Análisis de coyuntura, teoría e historia urbana, 103.

Echeverria, B. (1998). *Valor de uso y Utopía.* México: Siglo XXI.

Guevara, T., Raspall, T., & Zapata, M. C. (2011). Acceso al suelo de calidad para sectores populares. Balance de la Ley 341/964 y el Programa de Autogestión de la Vivienda. In M. Di Virigilio, H. Herzer, G. Merlinsky, & M. C. Rodríguez (Eds.), *La cuestión Urbana interrogada. Transformaciones urbanas, ambientales y políticas públicas en Argentina* (pp. 109–130). Buenos Aires: Café de las Ciudades.

Herzer, H. (Coord.). (2008). *Con el corazón mirando al sur. Transformaciones en el sur de la ciudad de Buenos Aires.* Buenos Aires: Espacio Editorial.

Holloway, J. (2010). *Crack Capitalism.* New York: Pluto Press.

Jameson, F. (1992). *Postmodernism: Or, the Cultural Logic of Late Capitalism.* Durham: Duke University Press.

Larraín, J. (2007). Ideología. Carlos Marx. Vol. 1. Santiago de Chile: LOM.

Larrain, J. (2010). *El concepto de ideología. Vol. 4. Postestructuralismo, postmodernismo y postmarxismo.* Santiago de Chile: LOM.

Lefebvre, H. (1969). *El derecho a la ciudad.* Barcelona: Ediciones península.

Lefebvre, H. (1992). *The Production of Space.* Hoboken, NJ: Wiley-Blackwell.

Lefebvre, H. (2003). *The Urban Revolution.* Minneapolis: University of Minnesota Press.

Lefebvre, H. (2013). *La producción del espacio.* Madrid: Capitan Swing.

León, E., Meave, K., & Ramos, A. (2009). Proyección territorial comunitaria en la Ciudad de México: El caso del Movimiento Urbano Popular. *Ciudades, 6*(9), 1–18.

Mançano, B. (2011). Territorio, teoría y política. In coordinado por Georgina Calderon y Efraín León (Eds.), *Descubriendo la espacialidad social desde América Latina.* México: Itaca.

Mayer, M. (2012). The Right to the City in Urban Social Movements. In R. Brenner, P. Marcus, & M. Mayer (Eds.), *Cities for People, not for Profit.* Oxford: Routledge.

MOI. (2012). ¡*Un grito en la calle*!. Buenos Aires: MOI.

Nicholls, W., Miller, B., & Beaumont, J. (2013). *Spaces of Contention. Spatialities and Social Movements*. Farnham: Ashgate.

Ostuni, F. (2008). Renovación urbana y sector inmobiliario: algunas reflexiones a partir de La Boca, Barracas y San Telmo. In H. Herzer (Coord.), *Con el corazón mirando al sur. Transformaciones en el sur de la ciudad de Buenos Aires*. Buenos Aires: Espacio Editorial, 78–90.

Ramirez, J. (2011, December). Restructuring Puerto Madero, Buenos Aires. *Portusplus-RETE Asociación para la colaboración entre Puertos y Ciudades*.

Rodriguez, M. C. (2004). Habitat, co-operativismo autogestionario y redefinición de las políticas públicas: buscando la nueva fábrica en los barrios de Buenos Aires. *Revista argumentos, 4*, 1–10.

Rodríguez, M. C. (2009). *Autogestión, políticas del habitat y transformación social*. Buenos Aires: Espacio Editorial.

Rodríguez, M. C., Bañuelos, C., & Mera, G. (2008). Intervención-no intervención: ciudad y políticas públicas en el proceso de renovación del Área Sur de la Ciudad de Buenos Aires. In H. Herzer (Ed.), *Con el corazón mirando al sur. Transformaciones en el sur de la ciudad de Buenos Aires*. Buenos Aires: Espacio Editorial.

Rodríguez, M. C., Di Virgilio, M. M., Arqueros Mejica, S., Rodríguez, M. F., & Zapata, M. C. (2015). *Contradicciones la constitución de la ciudad. Un análisis de los programas habitacionales de la ciudad de Buenos Aires en el periodo 2003–2013* (Documento de Trabajo No. 72). Buenos Aires: CLACSO-Instituto de Investigaciones Gino Germani.

Turner, J., & Fichter, R. (1972). *Freedom to Build: Dweller Control of the Housing Process*. New York: Macmillan.

Zapata, M. C. (2012). *El programa de autogestión de la vivienda: ¿una política habitacional habilitante del derecho a la ciudad?* Master thesis, Universidad de Buenos Aires.

Zapata, M. C. (2016). El rol de la institucionalidad publica en procesos autogestionarios de viviendas en Argentina. *Revista Economía, Sociedad y Territorio, XVI*, 229–264.

Zibechi, R. (2007). *Autonomías y emancipaciones. América Latina en movimiento*. Lima: Universidad Nacional Mayor de San Marcos.

Zizek, S. (2003). *Ideología, un mapa de la cuestión*. Buenos Aires: Fondo de Cultura Económica.

INDEX

N. M. Yip et al. (eds.),
Contested Cities and Urban Activism, The Contemporary City,
https://doi.org/10.1007/978-981-13-1730-9

T

U

The manufacturer's authorised representative in the EU is Springer Nature Customer Service Centre GmbH, Europaplatz 3, 69115 Heidelberg, Germany. If you have any concerns regarding our products, please contact ProductSafety@springernature.com

Printed and bound by CPI Group (UK) Ltd, Croydon, CR0 4YY
25/09/2026
02222895-0002